Acclaim for *How to Read a Tree* and Tristan Gooley

"We would be lucky to be lost in a forest with [Gooley]. Not just to find our way out—something he could surely help with—but to find our way in: to see what the trees are telling us about the Earth we all find ourselves a part of."

—*The Atlantic*

"This book will make wise naturalists out of us yet."

—*The Globe and Mail*

"An important book and a pleasure to read."

—Raynor Winn, author of *The Salt Path*

"Gooley interprets clues like a private investigator of the wilds, leaving no stone unturned. . . . For those inclined to solve mysteries written into the landscape, this author's lead is one they'll want to follow." **—*The Wall Street Journal***

"[Gooley] has become the global expert on natural navigation, finding his way around the world using nothing but natural clues and pointers." **—*The Daily Beast***

"Gooley is your man. . . . With unflappable practicality, he shares simple ways to understand your surroundings, whether you're beside a stream or on the open sea at night, without instruments." **—*Discover* magazine**

"Gooley's . . . true gift is in igniting curiosity and wonder about the world around us." **—*Shelf Awareness***

"Avid and budding outdoorspeople will appreciate Gooley's breadth of knowledge and accessible approach."
—*Publishers Weekly*

"Gooley, who has single-handedly been reviving natural navigation in this age of GPS, has the birdwatching skills of Bill Oddie and the deductive powers of Sherlock Holmes. He can make you feel that you've spent half your life walking about with your eyes only half-open." **—*The Telegraph***

"Most of us have lost our way in nature at one time or another. . . . Gooley can you lead you back to nature if you're willing to invest the time." **—*National Parks Traveler***

"Gooley's calm, contemplative authority on matters solar, lunar, and celestial establishes his guru credentials—but it's his revelations about the clues that lie scattered about the natural environment that really entrance: how puddles drying on paths, the shapes of sand dunes, the graininess of scree on the lee of a slope can all be enlisted to summon compass points to your horizon." **—*Time Out London***

HOW TO READ A
TREE

HOW TO READ A
TREE

Clues and Patterns
from Bark *to* Leaves

Learn to Navigate by Branches,
Locate Water with a Leaf,
and Unlock Other Secrets in Trees

TRISTAN GOOLEY

Illustrations by Neil Gower

THE EXPERIMENT

NEW YORK

The Experiment, LLC
220 East 23rd Street, Suite 600
New York, NY 10010-4658
theexperimentpublishing.com

THE EXPERIMENT and its colophon are registered trademarks of The
Experiment, LLC. Many of the designations used by manufacturers and sellers to
distinguish their products are claimed as trademarks. Where those designations
appear in this book and The Experiment was aware of a trademark claim, the
designations have been capitalized.

The Experiment's books are available at special discounts when purchased in bulk
for premiums and sales promotions as well as for fundraising or educational use. For
details, contact us at info@theexperimentpublishing.com.

Library of Congress Cataloging-in-Publication Data

Names: Gooley, Tristan, author. | Gower, Neil, illustrator.
Title: How to read a tree : clues and patterns from bark to leaves: learn
   to navigate by branches, locate water with a leaf, and unlock other
   secrets in trees / Tristan Gooley ; illustrated by Neil Gower.
Other titles: Clues and patterns from bark to leaves
Description: New York, NY : The Experiment, [2023] | Includes
   bibliographical references and index.
Identifiers: LCCN 2023006831 (print) | LCCN 2023006832 (ebook) | ISBN
   9781615199433 | ISBN 9781615199440 (ebook)
Subjects: LCSH: Outdoor recreation. | Orienteering. | Navigation. | Trees.
   | Handbooks and manuals.
Classification: LCC GV191.6 .G656 2023 (print) | LCC GV191.6 (ebook) |
   DDC 796.5--dc23/eng/20230223
LC record available at https://lccn.loc.gov/2023006831
LC ebook record available at https://lccn.loc.gov/2023006832

ISBN 978-1-61519-943-3
Ebook ISBN 978-1-61519-944-0

Cover design by Beth Bugler
Text design by Jack Dunnington
Author photograph by Jim Holden

Manufactured in the United States of America

First printing May 2023
10 9 8 7 6 5 4

To my godsons, Joey, Hector, and Jamie:
Happy Navigating!

Contents

The Art of Reading Trees

AN INTRODUCTION

Trees are keen to tell us so much. They'll tell us about the land, the water, the people, the animals, the weather, and time. And they will tell us about their lives, the good bits and bad. Trees tell a story, but only to those who know how to read it.

Over the years, I have enjoyed collecting every meaningful characteristic of trees that we can observe. It started with natural navigation and an obsession with the ways in which trees can make a compass—they grow bigger on their southern side, for example. This developed into a fascination with how trees make a map for us: The trees that grow by rivers are different species from those on hilltops. And this blossomed into a curiosity about the subtler clues, the patterns that hide in front of our eyes.

Do two trees ever appear identical? No, but why? Every small difference in a tree's size, shape, color, and pattern reveals something. Each time we pass a tree we can note a unique feature and read it as a clue to what that tree has experienced and what it reveals about the spot we stand on. A tree paints a picture of the local landscape.

The smallest details open bigger worlds. You notice that the leaves on a tree have a strong pale line down their middle and recall that this is a sign of water nearby. Moments later you see the river. Many trees that thrive near water, including willows, have that distinctive white rib on their leaves—they look like they have a stream down their middle.

My aim in this book is for us to immerse ourselves so deeply in the art of reading trees that we learn to find meaning where few would think to look. And once we have seen these things it's impossible to unsee them—trees never appear the same again. It's a joyous process.

We're about to meet hundreds of tree signs. I encourage you to go and look for them as this is the best way to become part of their story. It will help you read, remember, and enjoy them for the rest of your life.

1

The Magic Isn't
in the Name

THE ART OF READING TREES is about learning to recognize certain shapes and patterns and understand what they mean. It is not about species identification. The names of trees are a lot less important than many people think.

Individual species exclude people and tie us to certain places or regions—there are no native species common to both north and south temperate zones and possibly only one that Eurasia and North America share: the common juniper. There isn't a soul on the planet who can identify the species of most of the trees on Earth on sight. There never has been and never will be. It would take more than one lifetime to learn how to identify each willow species on sight, never mind that there are perhaps a hundred thousand other tree species. Recognizing tree *families* can be helpful, but individual species, not so much.

You will see I refer to common families, like oak, beech, pine, fir, spruce, and cherry. They are widespread. Most people can recognize a few of them and the others are easy to add. If you are totally new to trees and don't yet recognize any families, like oaks or pines, I have included some tips at the end of the book. Unless otherwise stated, we are considering the north temperate zone, which includes most of the populated parts of Europe, North America, and Asia.

In each case I am referring to broad traits within the families, not hard and fast rules that apply to every species or subspecies. If you can think of an exception, I congratulate you but hope you can appreciate that it is an exception that proves the general rule. A book that covered every such exception would be a dull book indeed and would quickly be returned to the tree pulp whence it came.

Some trees have many different names, and the "correct" one depends on which culture you ask. Indigenous peoples find remarkable meaning in plants but have little use for Latin. Whatever we call a tree, it can't change what we see or what that means. The fascination lies in discovering the global language of natural signs. I love the idea that we can recognize patterns in nature that someone on the other side of the world would also know, even if we don't speak a word of each other's language. Our ancestors' fluency in reading nature's signs must predate even the earliest of spoken languages by tens of thousands of years.

The word *magic* has more than one meaning. It can mean to perform tricks for entertainment. But it is also the possession of extraordinary powers, the ability to make things happen that would usually be impossible.

The roots of a tree will show us the way out of a wood, even when we don't know its name.

2

A Tree Is a Map

Where Conifers Call the Shots • Down in the Woods Today •
The Keys

I MADE MY WAY NORTH along the gentlest of rolling ridges in the mountains of the Sierra de las Nieves National Park in southern Spain. There was no path, but I picked dusty threads that twisted between rocks, gorse bushes, and thistles. The heat from the August sun rose off the land.

The sharp rocks meant I had to scan the ground, and every couple of minutes I'd pause and lift my eyes to take in the land around me. This is an old habit: When a path is difficult, we see too much ground, and when it is easy, we see too little. If you want a full picture of the land you move through, it helps to look down on good ground and up on bad terrain. But pause before looking up from a difficult path or your face will meet the rocks. When passing trees, tricky paths mean you see the roots and miss the canopies; easy paths mean you see whole trees, but miss the roots.

The scan paid off. Down in the gentlest of dips between hills I saw a green beacon, a clump of trees that didn't fit the pattern at all. I made my way down toward the greenery. Suddenly I could hear and see more birds, and some pale butterflies danced across my view. There was a slight change in the smell of the air. I took in slow, deep lungfuls. It wasn't a specific odor, just the familiar rich whiff of verdancy and decay. Then I noticed the animal trails start to funnel together and intertwine, like strands of a rope. Minutes later I was standing under a grove of magnificent walnut trees, the only ones for miles around. Near it there was a stone watering trough for goats, a confusion of their hoof prints in the wet mud around it.

The trees had signaled change: They had led all the animals, including me, to water.

Trees describe the land. If the trees change, they are telling us that something else has also changed: There has been a shift in the levels of water, light, wind, temperature, soil, disturbance, salt, human or animal activity. When we learn how to spot these changes, we have the keys we need to see the map the trees are making. We shall meet the keys soon, but first we will tune into two of the big broad changes we will see.

WHERE CONIFERS CALL THE SHOTS
After leaving the walnut grove, every large tree I saw for the rest of my walk in the Spanish mountains was a conifer. There is a good reason why.

A very long time ago, there wasn't much going on, but then evolution rolled up its sleeves. Algae appeared in the sea, then mosses and liverworts appeared on land. Soon, and by soon I mean a few hundred million years later, ferns and horsetails were spreading their simple fronds above the mosses.

Evolution is a genius at solving problems. It worked out that seeds meant you could give offspring a start in a different location and that led to most of the plants that grow today. Next it discovered that a woody trunk allowed you to stay above the competition for many seasons without starting at ground level again each year. Boom! Trees were born.

The earliest trees belong to the gymnosperm group of plants and include conifers. They bear their seeds in cones. About two hundred million years later another tree family evolved, the angiosperms or flowering plants, and this group includes most of the broadleaf trees. They are much more diverse in appearance than conifers but tend to have flowers that are easy to see and bear seeds in fruits. Most conifers are evergreens and most broadleaves are deciduous, losing and regrowing their leaves each year.

We can normally easily identify which of these two main groups we are looking at. If a tree has dark, needle-like foliage, it is almost certainly a conifer. If a tree has wide flat leaves and doesn't look like a conifer or a palm, it's very likely to be a broadleaf. (Palms have their own world that we will come back to.)

Conifers and broadleaves are in competition in many habitats and structural differences determine which group will succeed. The basic rule is that conifers are tougher: They can survive in many situations where broadleaves struggle. Evergreen conifers can photosynthesize all year round, even at very low levels, which means they do better than broadleaves in zones where the summers are cool and the sun is low. The farther from the equator we travel, the weaker the sun and the more likely it is that conifers will dominate. For example, we can expect to see more conifers in Canada and Scotland than in the US and England.*

Conifers have short, thin leaves that are better at conserving water, so they tolerate dry regions better than broadleaves. That is why I saw so many conifers on the dry Spanish mountainside. It is also why we will see a higher ratio of conifers in Mexico and Greece than in the US and England. But we can be more forensic about this.

If a large region has enough rain for broadleaf trees, but we don't see many, the water may be disappearing somehow. Sandy or rocky soils favor conifers, partly because the water drains away too quickly for broadleaves.

High ground tends to be drier than valleys, which is why you will sometimes find that conifers dominate the

*At even higher latitudes, as we approach the polar limits, it flips again and broadleaves reappear. At these extremes, trees can't maintain leaves throughout the year.

hillsides, but broadleaves line the river. Conifers are a darker green than broadleaf trees and this leads to interesting and colorful patterns in the landscape (conifers are mostly evergreens and need thick, tough skins and waxes on their leaves, which makes them appear darker). This is something you will have seen many times, but perhaps never noticed. It's very satisfying indeed when we understand why a fat ribbon of paler broadleaf trees marks a river's course. And this satisfaction makes us more likely to look for and spot it. We don't just see dark green and light green woodland, we understand it is a sign: We know the meaning of the color change and our brain likes that, rewarding us with the pleasant sensation that neuroscientists call dopamine, but we know sounds like "Ah!"

Plants have sap that carries water and nutrients from their roots to their higher parts, but the way this happens is widely misunderstood. The tree loses water from its leaves to the atmosphere through a process called transpiration. This leads to lower pressure in the vessels in the leaves higher up than in those at the bottom of the tree. The sap isn't pushed from below but is pulled toward the top of the tree by the lower pressure there. In kind climates this is a stable system, but it is delicate, and all plants have some vulnerability to freezing temperatures.

Broadleaves by the River, Conifers on Higher, Drier Ground

Even if a plant survives freezing, the process of thawing can introduce bubbles or "cavitations" into its vessels, which then block the pipes. Broadleaf trees have broad, open vessels that transport sap quickly and efficiently, but these larger vessels are particularly vulnerable to freezing. Conifers transport water from their roots using narrower structures called tracheids that are more resistant to cold temperatures (because the smaller bubbles quickly redissolve). If we look up from the bottom of a mountain, we can see the zone where broadleaves give way to conifers. It is never a perfectly straight line, but above that band broadleaves find it increasingly hard to survive and conifers outcompete them.

If a moist region stays warm all year, removing the risk of frozen sap, broadleaves are likely to do better than conifers. We see a lot more broadleaf trees than conifers in the tropics.

If you're wondering why all trees didn't evolve with the freeze-thaw resistant vessels that conifers have, the answers lie, as they so often do with evolution, in efficiency and survival. Broadleaves have a more efficient system, so they do very well *if they can survive.* But, as the saying goes, you've got to be in it to win it. Conifers are the tough but inefficient 4x4s of the road; broadleaves are the modern road car— much more efficient, but falls to pieces on tough terrain.

There's a couple of fun exceptions to the freezing vessel rule. Birch and maple are broadleaf trees that have devised an ingenious method for dealing with the problems created by frozen sap. They create positive pressure in their narrow vessels, which means they pump sap upward. This gets rid of any bubbles caused by freezing, effectively clearing the pipes in spring. It is how these species survive much farther north than we would expect. The boreal forests of Russia are a good example: They include many conifers, but large areas of birches, too. The positive pressure means these trees have sap that flows out of any cuts in the bark, making it easy to harvest and giving us birch and maple syrups.

Every time we notice broadleaves give way to conifers, we can assume that the environment has become tougher and ask ourselves how and why. The answer is likely to be temperature, soils, water, or a combination, and is part of the map the trees give us.

Spotting this change also shines a light on the psychology of perception. Ask someone to describe a landscape and they may include the word "trees" and not register the change in the woods before their eyes; ask the same person if there are different trees in that same landscape and suddenly the shift from broadleaves to conifers screams out. We have an extraordinary level of control over what we notice, but it is a choice: Nobody is standing next to us to ask these questions.

DOWN IN THE WOODS TODAY

Later in my Spanish micro-exploration, I ventured into a wood. It wasn't easy. The first ten minutes were slow-going as I had to pick my way through thorny bushes; they were only waist-high but determined to resist me. Next there were head-height hawthorns, then some trees I didn't recognize that rose a little taller— nettle trees, I believe—followed by some holm oaks twice my height. Finally I reached a forest of pines that towered above me.

Whenever we step into woodland there are certain patterns we can expect to see. The trees will get taller as you walk into the wood because those at the edge bear the brunt of the strong winds that buffet all exposed trees, which grow shorter as a result. The tallest trees tend to be near the middle of a wood.

The species will change as you walk into a wood, too. Trees that grow in the heart of woods are always different from those at the edges. Most trees follow one of two strategies: hare or tortoise. The hares are called "pioneers": They produce millions of tiny, often airborne seeds that land on any bare patches of earth. They start life quickly and grow rapidly. But the price they pay for this fast-out-of-the-blocks approach is that they don't invest in big strong trunks so there is a limit to how tall they can grow. Birches, willows, alders, and many poplars are good examples of pioneer trees.

The tortoises are known as "climax" trees and take a different approach. They produce much bigger seeds and play a slow-and-steady game because they know they will win in the long run. Oaks are a good example. We find pioneer trees at the edges of woods and in clearings, but climax trees in the older heart. If you walk into mature woodland, with trees that have high canopies, you can expect to pass through the shorter pioneers at the edge on your way to the taller climax trees.

Most pioneers appear lighter in color than climax trees and also cast less shade. Think of birches versus oaks. The birches have light bark, but they also let in much more light from the sky than oaks. This compounds the dimming effect as we walk into the woods:

The light levels drop a little as we pass among the pioneers at the edge but fall dramatically at the line where we meet the climax trees.

If you walk into a clearing with lots of pioneer trees, you are standing in a landscape in transition. Future generations will find the bigger trunks and deeper shade of the climax trees. The tortoises have won the race.

THE KEYS

It's time to narrow our focus and look for the clues the tree families are offering. Here are the major patterns to look for.

WET GROUND

Most species struggle if their roots are waterlogged, as it impedes gas exchange, but these families do well in wet soil: alders, willows, and poplars.

Ajay Tegala is one of the local rangers and a naturalist at Wicken Fen, a nature reserve in Cambridgeshire, in the East of England, and one of the most important wetlands in Europe, with more than nine thousand species of plants and animals. Ajay refers to the "tallest tree in the Fen," which is a backhanded compliment: There aren't many mighty trees in a habitat of peaty wetland. Ajay can see a single defiant poplar tree from anywhere in the reserve so he knows it well. I hear excitement in his voice when he talks about it.

Dry Ground

As we have seen, conifers tolerate dry conditions better than broadleaf trees. Among the broadleaves, these cope better than most with dry soil: maples, hawthorns, beeches, holly, and eucalyptuses.

I live on dry chalky soil and, for a fun micro-challenge, I set out from home to find the shortest line that would lead me past as many as possible of those trees. I forged a path that took me from a yew, past many beeches, a hawthorn, a pair of holly bushes and a field maple in less than ten minutes' walking. To add a eucalyptus, I would have had to walk for hours to find one in someone's garden—eucalyptus trees are native to Australia—but five out of six in ten minutes isn't bad. The same challenge on wet soil over clay or granite would be a long, difficult, and probably pointless task. I asked Ajay Tegala how he would find the same challenge in the wet peatland of Wicken Fen.

"It would be very hard! We don't have any yews on the whole reserve, and I'm pretty sure there are no beeches either. There are very few hollies. It would be an incredibly long walk to take in any more than just hawthorn and maple!"

Both Extremes

Unusually, the silver birch can cope with wet ground and moderate drought. I have so much respect for the silver birch: It's the tree that would be the last to grumble on a cold, wet family camping trip.

Lots of Light

Most trees show a preference for either lots of or little direct sunlight. As a general rule, conifers like lots of light, and broadleaves do well with some shade. Within each group there are hierarchies. Pines prefer more direct sunlight than firs, which like more than spruces, which need more than hemlocks.

Pines **F**eel the **S**un's **H**eat.
PFSH: pines, firs, spruces, hemlocks.

The following families thrive in bright, sunny environments: poplars, birches, willows, and most conifers, but especially pines and larch.

Many light-loving trees do well in the open. You will regularly see pines, poplars, birches, and willows from a distance. When they grow in woods, they do better on the bright south side. A line of pines on the south side of woodland is a common sight.

Shade-Tolerant

Trees that tolerate shade are tortoises and their tolerance is a big part of their strategy. A shade-tolerant tree can grow up slowly under light-loving trees and then, eventually, overtake them and cast shade on the competition. At that point it's pretty much game over and the tortoise wins: The latter can't cope with the shade. These trees tolerate shade well: beech, yew, holly, and hemlock.

Yews don't bother trying to grow taller than the canopy. They just shrug their shoulders and get on with life in the shade. Fair play to them.

Shade-tolerators thrive in the company of other shade-casting trees.

Exposure

Each tree has its own sensitivity to low or high temperatures.

As altitude increases, the average temperature drops and average wind speeds increase. As mentioned earlier, when we look from low to high up a mountain, we will see broadleaves give way to conifers, but also that all tree species grow shorter with increasing altitude. I refer to these two habits as the "tree altimeter."

On mountains there is an altitude at which even conifers find life tough and foresters stop growing trees commercially as the yield is too poor. This is where the neat-looking plantations stop. Conifers will survive above the plantations, but they are shorter and scruffier than lower down the mountain. We start to see gaps between the trees.

Conifers are vulnerable to wind injury so it is common to see conifers on mountainsides that have outcompeted their broadleaved cousins but look battered by the experience. The dwarfed and malformed conifers that cling to life in these high cold zones are called *krummholz*, a German word meaning "twisted wood." A little higher, the climate is too brutal for any trees and they give up at an altitude called the tree line.

In hotter climates, the three main cedars, lebanon, deodar, and atlas, all cope well with warm mountainous habitats.

Soil

The first chapter of *Sylva*, John Evelyn's landmark seventeenth-century book about trees, is peppered with references to the soil that trees grow in, but even in the twenty-first century this remains a young science with many gaps. Fortunately, some bold patterns are easy for us to spot.

There is rich and poor soil. Some soil is rich in the nutrients that plants need for healthy growth, including vital minerals like nitrates; poor soil lacks these essential chemicals.

Ash trees like moist but not wet soil and are fussy about nutrients: They need richer soil than most. Ashes are more common low in a valley than high up. In a river valley, there is normally a band of moist soil near the river, but not so close that it is waterlogged, and it is rich in nutrients that have flowed down from the high slopes. This is the ash's favorite spot.

Walnut trees like deep, nutrient-rich soil. The ones I met in the Spanish mountains had found the only spot in the area where life for them was possible. Water and nutrients had gathered in the deeper soil in the dip between two small peaks. It gave the walnuts exactly what they needed. If I had picked up a walnut and thrown it in any direction, it would have landed on soil that was too dry, too thin, and too poor for a walnut tree to grow.

Elms also like nutrient-rich soil.

Major fluctuations in the pH of soil—its acidity or alkalinity—can radically change which trees we will see. This consideration overlaps with nutrient richness because acidic soils are usually low in nutrients.

Alders and willows are likely to do well in wet ground, unless it is acidic, with perhaps peaty soil where downy birch will do better. Conifers cope well up to a point with acidic soil.

Towns

The urban environment is tough for trees, with heavy footfall and motor traffic, but there are less obvious stresses, too. It is warmer and drier than the surrounding area; there may be de-icing salt, dog mess, and a long line of people wanting to dig up the world.

The London plane has been planted in towns and cities around the world because its roots tolerate compaction of the soil and its bark sheds regularly, allowing it to withstand more pollution than many. The sycamore is a member of the maple family that copes well with the stresses of town life, too well perhaps: It has a reputation for sprouting up uninvited in gardens and parks.

I once visited the coastal Devon town of Budleigh Salterton, in southwest England, to give a talk in a church. I parked the car and ventured out to find the venue, but all I could remember was the name of the church and the part of town it was in. I ran around looking for yews, discovered some lining a residential street, then peered through a gap and found the venue just in time. Yews have been planted for centuries in many churchyards and other significant sites in towns. (In rural areas, yews are a sign of few grazing animals as their toxicity means the two don't mix.)

Trees don't form straight rows in natural environments. Even those that line a river will show curves that reflect the bends. It follows that any straight line of trees is a sign that humans are behind it. The most obvious are the formal avenues of trees leading to something grand at the end, but there are many more interesting examples.

Lombardy poplars are often planted in a line that marks the edge of a property, village, or farm. They are so easy to recognize once you know them, standing taller than the other trees in the landscape, with thin branches that reach for the sky. With practice it becomes instinctive to spot their forms, and I regularly use them to identify the location of a hidden village. The Lombardy is a member of the water-loving poplar family, so it is often a double clue: civilization next to water.

Lombardy Poplars

It is so satisfying when we connect the dots in a land-scape. The other day I set myself the challenge of descending a Sussex hill and finding a village, using only the trees for guidance. At the foothills of the northern scarp, I found ashes thriving in the rich, moist soil; a little farther on willows lined a stream. The water led me to the village, and I knew I had arrived when the horizon was broken by a proud line of Lombardy poplars.

DISTURBANCE

All plants are sensitive to disturbance. If the land is ravaged by storms, fire, water, human clearance, or heavy use, certain trees give up on it for long periods, while others are happy to start again as soon as the drama is over. The following families are keen colonizers, springing up in disturbed areas—if you see lots of young ones, it is a sign of a major disturbance: willows, alders, larches, birches, hawthorns.

All of these trees are pioneers, the hares, winning in the short run, but most will be gone within a century, having been replaced by the climax tortoises. This means they form a particular sort of map. They hint at motion and upheaval and tell us of a recent major change in the landscape. We should look for the cause.

Larches are colonizing conifers and look quite different from most other conifers throughout the seasons. They have distinctive paler foliage in summer and, unusually for conifers, they are deciduous, losing their needles in winter. Larches spring up in places where humans try to cut ribbons through the trees and make a useful map, marking regularly used forestry tracks through coniferous woodland. Look from a local peak and you will see their pale lines snaking through the darker trees, marking well-worn vehicle tracks. Often you can spot a bigger clump of larches surrounding a depot or a spot where forestry work is going on.

In areas that are prone to wildfires, a different type of competition is taking place. No trees find fire easy, but some have evolved to endure it better than others and tend, over time, to outcompete the vulnerable. For example, the Douglas fir sees off most of the competition in the fire-prone regions of the Pacific Northwest.

There are interesting patterns on the charred trunks of pines on many wild, mountainous landscapes. In places like La Palma, in the Canary Islands, the trees are tough, coping with dry, rocky terrain and altitude, as well as being fire-resistant. As a wildfire moves through the trees, it will char and scar one side more than another. If you take the time to note the darker sides of the pines, this consistent trend can act as a compass for natural navigation.

THE COAST

Before you pick up the first whiff of sea air, there may already be enough salt in the air to kill off many plants. The desiccating effect of salt reaches 12 miles (20 km) inland. By the time we can see the sea, most inland plants will have given way to marine-tolerant species or, at the very least, they will show signs of struggle in their leaves. When we can feel the spray on our faces, few plants can survive: Only the hardcore specialists remain. A few unbelievably hardy lower species, like sea kale, a cabbage-like plant, can live on a stony beach in the splash zone, but it's not a home for the sensitive. No trees need the salt levels provided at the coast, but a few can tolerate it.

Sycamores do surprisingly well near the sea; they have thick, waxy leaves and roots that resist the salt. I remember walking along a coastal path in Pembrokeshire, Wales, and as I approached a headland, I walked for half an hour before I saw any trees. Then I spotted a clump of sycamores, sculpted by salty winds and clearly battered, but proud and defiant. On the sea side, their leaves were brown and crinkled, "burned" by the salt.

Nearer my home, the only trees that survive so close to the sea tend to be tamarisks. I find any trees that beat the odds beautiful and beguiling. The other day I stood on West Wittering beach in Sussex, South East England, leaning backward into an onshore gale, as I admired a line of tamarisks. It was September, and the fierce equinoctial wind was lifting sand off the beach and keeping swimmers, even surfers, out of the water. And yet there was the tamarisk, holding its line as salt spray whisked past its spikes of pale pink flowers.

One reason why so many holiday resorts look similar is the often-uninspiring architecture. Another is that only certain trees can tolerate the things we like on holiday: heat, sea, and sand. The palm tree has become a marketing symbol: It subliminally suggests we will get sun, sea, and sand. The palm is a tough and peculiar tree. It has followed its own evolutionary path, closer to grasses than most other trees, which has allowed it to survive on beaches and in brochures.

Coconut palms lean toward the sea so that they can drop their seeds, the coconuts, into the water ready to float off for a new life farther along the coast or on another island. Most beaches experience "sea breezes," cool winds that flow from sea to land. The coconut palm's trunk leans toward the sea but its top is regularly blown in the opposite direction by the breeze, giving it that trademark shape: the trunk leaning out to sea, the top bending in the opposite direction.

The sea brings harsh salty winds, but it's not all bad news for nature. It also brings warmer air in winter and cooler in summer. Palms hate frost and thrive near the coast, surviving close to the sea even in cool, temperate climates.

In a few places, the oceanic climate reaches just far enough inland to create a unique and rare biome called a "temperate rain forest." When the warm, very moist air travels inland it loses much of its salt but retains most of its moisture and mild temperature. This guarantees high humidity and low variation in temperature. The west coasts of North America and Europe, including parts of the UK and much of Ireland, have pockets or strips of temperate rain forest. I once spent a wet happy day in one in Devon and it's easy to see why they're called rain forests: They're like mild-mannered jungles—the lush verdancy is intense.

At the end of my time in the Sierra de las Nieves in Spain, I drove off the mountains, the open window letting in the scent of the pine forests. The road twisted and descended. The pines changed to oaks that continued all the way down to the coast, where I parked the car and walked to the beach. The last trees I passed before I plunged into the water were palms.

3

The Shapes We See

The Risk-Takers • Little or Large • Cones and Balls • The Graffiti Stays •
The Parasol Effect • Skinny Shade, Fattening Sunshine •
How Many Layers Does a Tree Need? • Out and Then Down

O N A WARM APRIL EVENING, I walked home over the hills after dinner with friends at a local pub. The sun had set an hour earlier, the air was warm, and there were few clouds. Without invitation, the elements put on a show. The bright stars of Orion teamed up with Mars, and the faintest sliver of a moon appeared to the west. The last light was fading as the final pink and orange hues hung behind the trees. The woodland drew a solid dark line above the horizon, but the silhouettes of individual trees I passed had more character.

If you had been with me, I'm sure you would have spotted the spire figure of a lone tree and recognized it as a conifer. And the more globular shape we passed a few minutes later would proudly announce that it was a broadleaf, in this case, an oak. But there, at the edge of the village of Slindon, backlit by the last rays of the sun and the first of the villagers' lightbulbs, is a different form. The pendulous, drooping branches of a pair of birches hang, almost sad, different from the first two forms we saw. Samuel Taylor Coleridge called the birch "The Lady of the Woods" and perhaps there is something feminine about the soft, thin branches that flow downward.

Three very different tree shapes in the space of a thousand feet. How many basic forms are there? A few, a hundred, an infinite number? There are thousands of different species, but a 1978 academic study of tree forms concluded that there were only twenty-five basic shapes. It's a nice idea, but each scientist could argue a different figure. What is more important and much more interesting are the reasons for the shapes.

The trees we see reflect the local world around them and us. This is the selective pressure of the environment, and the rules of the game are simple: If you can't survive, you don't survive. This applies a filter, and we see only the winners, but that poses an interesting question: Why don't we see lots of identical-looking trees?

There are three reasons.

First, the kinder the soil and climate, the less harsh the filter and the more different species can survive. This gives us a basic sign: If you see lots of different tree shapes in a landscape, the environment is kind and easy. You will see plenty of humans, animals, and smaller plants, too.

Second, each tree leads a different life from its neighbors and its shape reflects that. The two birches at the edge of the village were obviously the same species, but they looked different and asymmetrical, with branches falling away from each other. Climate, weather, light, water, soil, competition, disturbance, animals, and fungi can all change a tree's shape. The birches fell away from each other because the older of the two grew toward the southern light, but the younger leaned away from the shade of its older sibling. This is a pattern you will spot whenever you see any pair of trees growing close to each other—the older one grows toward the southern light, the younger toward the only remaining light, which means growing away from its neighbor.

Third, time plays an important role. We will not see the same things or possibly even the same trees if we come back a couple of decades later. If I'm lucky enough to have grandchildren and they follow the same walk, each of the Slindon trees will look different. There is a good chance that the birches have disappeared: A birch's lifespan is similar to that of a human.

Every tree we see reflects these three influences: genetic, environment, and time. Once we learn to spot these sculpting forces, their impressions become stories and the stories hold meaning. Let's look at examples of each in turn, starting with the genetic.

What makes trees so good at surviving in nature? What is their secret? If we can answer this, we can start to make sense of the shapes we see. The process of elimination helps: If we find the one thing they all share, it must hold an important clue.

It's clearly not the color or pattern of the leaves, bark, or roots, which vary enormously across the species. It's not the way the trees reproduce: Conifers and broadleaves have quite different approaches. The one thing that all trees share is a bit of height on a trunk that lasts through the seasons and years.

Height is critical: The taller a tree is, the more likely it will receive plenty of light. Surely, then, the trees that can grow tallest will outcompete all the others and we will be surrounded by only very tall trees. That's not what we see. Growing tall requires huge amounts of energy and means transporting tons of water to great heights. It also makes trees more vulnerable to the wind and weak soil. A bit of moderation is needed, but if that solves the challenge, why aren't we surrounded by lots of fairly tall trees? Because of the "tower building problem."

Imagine a bored, rich, and slightly mad friend invites you and one other friend over to play a game. You are each given a box of small wooden bricks and told you have fifteen minutes to build the highest wooden tower you can on a table. If your tower topples over, you lose instantly. The winner will receive a thousand pounds, the loser nothing. You are sent into different rooms and can't see each other. How do you win?

Once you are a few minutes into the challenge, you start to see that this is as much a test of character and strategy as of skill. Do you stop as soon as you reach a respectable height, or do you keep trying to build more levels? As the minutes tick by you realize that one more layer of bricks may bring the whole thing down and you'll lose. But playing safe is risky, too. If your friend thinks you'll play it safe, they may take a bit more risk and beat you. There are no prizes for coming in second.

Trees also face a strategy dilemma. If they invest all that energy in growing to a great height but don't win the prize, lots of light, they lose. And, since nature can be an evil genius as well as a friend, losing means death.

So, if you look at a woodland from a little distance, you will often see one or two trees a little higher than all of the others. These are the risk-takers, the trees that have gone for bust, building an extra layer of wooden bricks. They are happy to risk a storm's battering for the grand prize of all that sunlight. In every woodland there will be trees that show a little more appetite than their neighbors for risk. On my walks, I notice that the beeches are happier to take on the dangerous winds than the oaks and tend to poke up above them.

If you're going to play the game of trying to grow tall for lots of light, you have to play to win. You've got to reach the best height you can without failing structurally or falling over. But what if there is another way of playing?

When I was about eleven years old there was a brave overweight kid at our school. I'll never forget him. We used to have to go on cross-country runs about once a week in winter and most of us didn't really relish the effort required to run for an hour in the cold and wet. We used to grumble a bit and then get on with it. But that kid, I'll call him Jake, decided that this was a silly game and he didn't want to play. He had to come on the run, it was compulsory, and he'd get into more trouble than he wanted if he didn't. But Jake chose to go at his own pace. And it was slow. A bit of unenthusiastic jogging and then he'd walk a bit, stop for a second or two, and jog for a couple more minutes.

At the start of each run, we'd glance back and watch Jake disappearing from view, each of us dumbstruck by his willingness to stand out in this deliberate way. When we finished the circuit, we'd rest, hands on knees, panting steam for a bit and chatting about the muddy ditches we'd struggled over. After a few minutes, we'd look back, almost excited to see how long it would be before Jake appeared. Sometimes we'd be getting out of the shower by the time he jogged round the last bend, always smiling, often laughing, but perhaps sad. He'd usually get a round of applause, which irritated the teachers enormously—and made us clap louder. Jake did not conform, which I would never have had the courage to do at that age or in that way.

There are nonconformists in nature, too. What if, instead of trying to win the same game as everyone else, you change the game? What if the aim isn't to grow tall to get as much light as possible but to make the most of a little? Suddenly there is no need to beat all the other trees in the great race to the sky.

Some trees don't grow tall, ever: They stay low, growing only a little higher than the average adult. As I write this, I could reach through a window and touch the leaf of a mature hazel that is barely taller than me. It is growing in the shade of several beeches that tower 100 feet (30 m) above us. It is a Jake of trees and I hear it laughing at the over-eagerness of the taller trees. To paraphrase an old saying, "A clever tree solves a problem. A wise tree avoids it."

In many situations, like disagreements with friends or partners, compromise can be the best answer, but in nature it is often suicidal. In the height and light game, the worst possible thing a tree can do is use lots of energy to grow half the height of the tall trees and stop. It would run out of energy quickly because it would use lots and harvest little. Trees can't compromise with height. This is why we will find lots of tall trees and plenty of short ones, but very little in between. The short ones are typically about 8 feet (2.5 m), a bit taller than the average adult human, and the tall ones vary, but can easily grow to over 100 feet (30 m). If we see trees between the "little and large," they are most likely young large trees. In a natural setting, mature trees tend to be one or the other.

One more reason that small trees do better than medium: There is actually more useful light near the forest floor than halfway up the trees. It is a simple optical effect: Any light that sneaks through a small gap in the high canopy forms a cone shape, tiny at the top and much broader at ground level. As the sun moves over the canopy, the cones move over the ground. This means that there are some broad, longer-lasting patches of weak light near the ground, which is more dependable than the brief snatches of bright light that might be found halfway up the tall trees.

Little or Large

I should clarify something here. I personalize the trees' choices, because it is much easier to explain some concepts, like strategies, if we imagine ourselves in the game. Obviously, trees aren't thinking or strategizing in the way we might: Evolution has forced the choice long before the tree started life. The decision is hard-wired into the genes of each species: A tree is programmed to be either short or tall before its first leaf appears. If you're curious about how this happens, the evolutionary mechanism is straightforward.

Let's say there is an early species of tree and random mutations mean that one year it produces three seeds with different genes: One is a very tall version of the same tree, one is a very short version and one is a medium version. All three of those seeds land in good soil and germinate. But only the tall and the short ones live long enough to produce their own seeds. The medium tree has grown too tall to live in shade, so it dies and the "medium" gene dies with it. So, the next generation consists of very tall and very short trees. In this example it has taken two generations. In nature it may take thousands of years, but the effect is the same: Evolution kills off bad strategies.

CONES AND BALLS

The same evolutionary pressures that force a tree to be short or tall are at work in lots of other areas, too.

Conifers tend to be conical and most broadleaves are more rounded. Conifers have evolved to survive in high latitudes and to keep their leaves through winter, which may mean coping with a lot of snow. Snow slips off skinny trees with branches that flow downward; it builds up, then breaks the limbs of broad trees with flatter branches. A tall thin shape also helps harvest light from a low sun that barely climbs into the sky. In hot, dry regions, the same conical shape helps reduce the radiating heat near the middle of the day, when the sun is highest in the sky.

Trees are controlled by their summits. The highest growing part, the top of each tree, is called the apical bud. It releases hormones, chemicals called auxins, which travel down the trunk and dictate how every other main branch in the tree grows.

The way in which each species does this varies, but some general trends are easy to spot. Conifers have dictators at the top: Their auxins send a strong message to all the lower branches that they must grow slowly. That is why most conifers are tall and thin. The lowest branches have been growing slowly for longest, which is why the base is wider than the top and many conifers have a spire shape.

Broadleaves have weaker, less domineering bosses. The apical bud in these trees sends a softer message: It's fine to grow, but try not to grow faster than the top, and it's OK to spread out a bit. That is why oaks, beeches, and most other broadleaf trees have a more rounded shape than conifers.

Strong top: tall skinny trees, including conifers.
Weak top: rounded canopies, including oaks.

This all works fine until there's a coup. Cutting the top off a tree—a storm, a gardener, or an animal may be responsible—removes the apical bud boss. The auxins stop flowing down the trunk and this takes the brakes off the lower branches: They start to grow more quickly and new ones appear. This is an effective survival mechanism: It allows a tree to change the strategy and start growing again after any calamity.

Decapitation causes some interesting patterns—it's why hedges are so dense: Each pruning leads to more and more tiny branches. It's also how commercial growers make Christmas trees bushier and less spindly. We will return to these patterns when we look at branches more closely, but for now, try to spot a few different tree shapes and work out what sort of boss is at the top of the tree: domineering or easygoing.

THE GRAFFITI STAYS

It's time to change the way we think. If not a myth exactly, there is widespread misunderstanding about how trees grow. We need to sort that out if we want to understand the shapes we see.

Find a tree with a large low branch that you can just touch standing on tiptoe. By stretching as high as you can, your fingernail brushes the bark on the underside of the branch. Here's a question: If you come back in five years, will you still be able to touch that branch? (We're assuming you haven't grown, shrunk, changed shoes, or done anything else to confuse the experiment. We're trying to work out if that branch will be lower, higher, or the same height off the ground when you return.)

Perhaps you had the pleasure as a child of growing sunflowers, or some other miraculously fast-growing plants, from seeds. We're used to watching the seedling emerge and work its wriggling way upward. We can almost see the motion: Each week the tiny plant gets noticeably taller. And perhaps you have watched those videos of the same thing happening in quick time. These are interesting experiences, but they put a misleading idea in our minds about how trees grow.

Trees have two types of growth: primary and secondary. The primary growth is exactly like that of the sunflowers. A bud grows upward, forming a green stem. But once a stem has appeared, trees form bark and a different type of growth continues. Secondary growth is the fattening of existing bark-covered trunk and branches. Here's the key thing: Once bark has appeared, that part of a tree trunk is not growing upward any more. It gets fatter, but not taller. The same principle is at work in the branches: They continue to grow longer at the tip, but the parts nearer the trunk become fatter—they don't move outward.

The tallest tip of the trunk, the apical bud, still moves upward and gets taller with primary growth, but the lower parts don't. If you scratched a line on the bark, it wouldn't get higher each year. I don't encourage anyone to do that, as it is harmful to the tree, but you will have noticed that scratched graffiti on trees doesn't move up the tree, even after a decade. If it did, we would see "Leo 4 Gemma" high above our heads, but we don't. We see it at the same height it was when lovestruck Leo wrote it a decade earlier.

To answer our question: Yes, you would be able to touch that low branch in five years. In fact, it would be easier because your branch would be fatter, but no higher, meaning that the lowest part would actually be lower.

THE PARASOL EFFECT

The shape of each tree reveals its strategy. If a tree is going all out to win the height race, it will inevitably have lots of lower branches that are either in the shade of its higher branches, or soon will be.

These trees have a problem. They need every joule of energy they can muster to fight on up to the top of the canopy, but they are now carrying lots of shaded lower branches that aren't helping their cause. We know that these lower branches are stuck at that level: They'll never rise above the canopy. That is fine if you're a tree that is happy in shade, but tall trees are not. There's a neat solution: They shed those branches.

Trees that like lots of direct sunlight, like pines, grow tall and keep their upper branches but shed their lower ones. It gives them a "top heavy" appearance, which I think of as the Parasol Effect. The pines display this quite dramatically, but it can be seen in varying degrees in most species. As I look out of my cabin window, I can see the beeches overgrowing almost everything else, and they have few branches near the ground. A handful of much shorter species, including a couple of hawthorns and hazels, have kept their lower branches.

If you look at the outline of conifers, you will see this effect across the species. You'll recall that the pines like more sunlight than the firs, the firs more than the spruces, and spruces more than hemlocks—Pines Feel the Sun's Heat. And that is the rough pattern we see in their silhouettes: Hemlocks have more low growth than spruces, which have more than firs, which have more than the parasol pines.

Pine   Fir   Spruce   Hemlock

I walk the dogs regularly, and on one route we pass a western hemlock early on. Its leaves have a gorgeous grapefruit scent when rubbed between the fingers. I enjoy the occasional crush and sniff. Twenty minutes later, in a clearing near the top of a hill, we meet a Scotch pine standing tall and proud. Pine needles also have an interesting smell that brings a familiar sense of wilderness on a breeze—it's lemony and pleasant but with astringent hints, almost medicinal. (And it may be medicinal: It certainly kills many of the trees' pathogen enemies and research shows that it does us some good, too.) I like the idea of plucking some of those needles, crushing them in my hand and breathing in deeply, but it's not practical. The nearest living needles are 50 feet (15 m) above my head.

As a rule, if we can reach the leaves of a mature tree, it is a shade-tolerant species. I remember this with the line,

"Low leaves, low sun."

SKINNY SHADE, FATTENING SUNSHINE

Most of the trends we have been looking at so far are predetermined and hereditary: They lie in the genes. A fir tree will never look like an oak, whatever nature throws at it. However, plenty of patterns are governed by the environment, and sunlight is a major influence.

A tree doesn't know exactly how high it needs to grow to rise above its competitors, and it would be risky for it to grow any higher than necessary. The trees tackle this challenge with another simple solution—they sense and react to light levels. Growing upward is only about getting to the light: Once the tree senses bright light, it doesn't need to gain any more height, so it tweaks the plan.

For as long as the topmost tip, the apical bud, is in shade, it continues sending chemical messages to grow taller quickly and the lower branches are held back. As soon as it senses it's in full sunshine, it changes the message: The tree slows its upward charge and the branches spread out. We don't see this happen in real time, but the cumulative effects are conspicuous.

Pick any tree species you see regularly. You will soon spot how it tends to grow taller and skinnier when in woodland or other shady places, but shorter and broader in open well-lit situations. Foresters plant trees in tight rows because that makes the most of these effects. The trunk is the part of a tree with commercial value; the side branches are often in the way of the timber harvester. It is a little ironic that you raise the best tree crop by planting them in a crowded way, which means they get less of the light they need. But it works: They shoot up as tall, straight stems with minimum side growth.

This is also why you'll see trees expanding into the new gaps created by storms or clearances in woods. The sudden spike in light levels changes the way they grow, slowing their upward charge and increasing their sideways growth.

Sometimes you will come across a tree that seems to break this rule, perhaps a tall, skinny oak in the open, or a short, broad one in a wood. It is a clue that the landscape has changed. The thin oak in the open grew up surrounded by other trees that have since disappeared. The fat one in the woods had that patch to itself for many years but is now surrounded by faster-growing trees.

HOW MANY LAYERS DOES A TREE NEED?

Is it possible to have looked at something a thousand times and never seen it? Yes, and it's time for us to explore a good example.

It's tempting to think that leaves in full, bright sunlight will do twice as well as leaves in half that light, but actually, most leaves are working as hard as they can in about 20 percent full sunlight. This is odd, given how much effort the tree makes to push its leaves into the light, but if the top leaves are in full light, those a fraction lower may already be in partial shade for much of the day.

Trees that have evolved to grow in bright, open spaces, like birches and hawthorns, will receive lots of light at lots of levels. Trees that have evolved to grow up through a canopy, like most of the taller trees, will get little light until they break out and then only near their top. These are different light environments, and it would be strange if the two tree types used the same shape strategy. They don't.

Multilayer   *Monolayer*

Trees tend to be either "multilayer" or "monolayer." Trees that specialize in growing up in shade, then reaching the top of a canopy, like beeches, have a flatter structure. They put out most of their branches at a similar height, in one, "mono," layer. Trees, like birches, that grow in the open put out branches at many different heights, in "multi" layers.

When most people start to look for this effect, every tree seems multilayer. The reason is that we see so many trees in an open environment and, as we saw in the "fattening sunshine" effect, they adapt to become more rounded in bright spaces. But walk into a broadleaf wood and look up: You'll see plenty of monolayer trees. If most of the branches are so high that they'd be hard to hit with a stone, you're looking at a monolayer tree. If the branches spread from top to bottom and you could throw a stone over many, it's a multilayer.

It is a little hard to visualize how and why evolution created these two shapes. It gets a little easier if we imagine swapping sunlight for drops of water.

A couple of months ago, we noticed water dripping through our kitchen ceiling. My heart sank and I ran upstairs to the cupboard I suspected concealed the problem. Sure enough, water was dripping out of a valve near the bottom of the hot water tank.

Fortunately, the drip was slow, so I wedged a bowl under the valve to catch the water. Then, after fiddling about and failing to make any difference, I called Tom, our friendly heating engineer. (He's quite a character: He works on domestic heating systems alongside doing a PhD in atomic energy metallurgy, but that's a story for another time.) Tom promised he'd get there as soon as he could, but it wouldn't be for twenty-four hours at least.

"That's okay," I said. "It's only a slow drip and I can probably catch it all in a bowl." I allowed a little urgency into my voice.

For the next six hours I checked the cupboard frequently. The small bowl had filled so slowly that I hardly needed to empty it. But things changed. The rate of dripping increased. I texted Tom as the bowl filled quickly. My goal for the hours before he arrived was simple: I had to find a way to stop water flowing down from the valve to the floor of the cupboard. If I could do that, none would drip into the kitchen. I couldn't fit a bigger bowl into the original space between the pipes so I added another bowl of the same size and slid it in just below the first. By the time Tom arrived, there were four bowls in the cupboard, stacked one below another, allowing the drips to cascade downward and giving me time to swap the odd one out.

And this, bizarrely, is what leaves are trying to do with light. If we think of the light arriving in drips from above, the leaves are there to stop it reaching the ground: That would be wasted light, and nature abhors waste. If there isn't much light, a single bowl near the top does the job. But if more light arrives than one level can harvest, it makes sense to grow another level lower down.

In shady woods, the single layer near the top of the canopy catches all the light. In bright, open areas, we can think of the light cascading down the many levels of the birches. And the extra layers work doubly well when you've got light coming in from the side, too.

OUT AND THEN DOWN

We've now looked at some of the ways that genes change the shape of trees and also how they are sculpted by the environment—nature and nurture. The third area for us to consider is time.

Trees grow larger with time, but they also change form, especially when they're old. Most turn more ragged and asymmetrical in their dotage. In many species, the boss at the top mellows with age. Pines start life with discipline, good symmetry, and a clean pyramidal shape. In later life, they care less for rules and wear a more bohemian style. A tree may have a thin top and a disciplined shape in middle age, but there comes a time when the apical bud weakens, exerting less control, which leads in some trees to a flattening top. Yews grow upward when young, but spread widely in maturity, and you'll see the effect strongly in some older pines, too.

Once the apical bud weakens, the lower branches start to grow more vigorously, leading to a broader canopy shape. It also means a tree that started life as a monolayer will tend toward growing more layers with age. Most monolayers creep toward a multilayer appearance when they're old. I have an odd way of thinking about this broadening effect. The top bud is a strict, mean grandparent; the lower branches are restless children who want to go outside and play. The grandparent scolds the children: "No, you can't go outside. It's wet, and you'll get dirty and mess up the carpets!" But the children bide their time and eventually the grumpy old grandparent grows tired and nods off in their rocking chair. The kids dash out of doors.

When a tree is very old, a "veteran," the top may die long before the lower parts. Dead branches poke out of the top, with green healthy branches below, a process known as "retrenchment" or "growing downward."

Each species displays its own combination of these effects, some stronger than others. Some, like ancient oaks, have dead top branches that poke out with such a striking effect that it has led to a nickname: They're called "stag-headed" trees.

4

The Missing Branches

Thick and Thin • Empty Trees • Up, Down, and Up Again •
Missing Branches • Southern Eyes • Defender Branches • Plan B •
A Trunk-Shoot Compass • Escapes, Avenues, and Islands •
Opposite or Alternate • Zigzag • Orders • The Uneven Collar •
Pointing to Pioneers • Witch's Broom • A Little Too Friendly

BRANCHES HAVE THEIR OWN silent language. They can half fill our landscape and still remain inconspicuous. The next time you spot a tree you haven't seen before, turn your back on it. Now describe its branches in as much detail as possible, without peeping. Did you struggle?

In 1833 six agricultural laborers met under a sycamore in the village of Tolpuddle, Dorset, in southwest England, and agreed to make a stand against deteriorating wages and rights. They were arrested for swearing a secret oath, sentenced to seven years' penal labor, and transported to Botany Bay, Australia.

There was a mass outcry. Eight hundred thousand people signed a petition, and the seven were pardoned after three years as sheep farmers Down Under. It was a pivotal moment in the birth of the trade-union movement, and the seven are remembered as the Tolpuddle martyrs. The tree survives to this day and is now about 340 years old.

A few weeks ago, I was walking the dogs through the woods in the low light before dawn, when one made a strange noise behind me. It was our tiny runt Jack Russell—she's a bit neurotic. I turned to check she was all right. Stupidly, I didn't stop walking and my eyes were soon tickled by the low foliage of a hazel tree. I started to rub them but, having failed to learn my lesson, I continued walking blind. The next moment I noticed the light change and, squinting through watery eyes, I saw a dark shape. I ducked so suddenly that my knee hit a rock on the ground and I passed just under the jutting low branch of a large sycamore. I escaped with a scratch and a little blood, but no serious harm.

The branch I ducked below grows on a tree that is a similar size to that of the Tolpuddle martyrs. My tree will be remembered for nothing, but it raises a question. How was it that one large sycamore had a branch low enough to force me to the ground, while another of a similar age had enough room beneath it for six people to hold a meeting?

There is always a good reason for the height and position of the branches we see, and if only I had looked where I was going instead of pandering to my over-precious dog, I would have enjoyed reading the branches on that sycamore. In this chapter we'll discover the signs in the lines of the branches. We will start with the effects that are easiest to recognize and move through to more challenging trends.

Branches are thicker near the trunk and thinner at their extremes, which is obvious when we think about it, but we rarely notice it. The last time we appreciated this may have been when climbing trees as a child: The farther we crawl from the trunk, the greater the risk that the branch fails and we fall. In a general sense we all still know this, but the tapering trend varies across the species and is unique to each one. I find the following thought experiment helps to make these differences shine out. Imagine making a ring with your thumb and index finger and seeing how far along a branch you could slide it, from tip to trunk. Ignore any offshoot branches that may get in the way.

Once you start to look for this tapering, you'll notice how much it varies and how much more pronounced it is in trees that grow in the open, like the pioneer trees. Trees that have evolved to cope on their own are exposed to strong winds and their branches taper to the thinnest lines. Wires and whips are common at the end of pioneer tree branches. Birches take this to the extreme: Their branches are so thin at the ends that it seems electricity would struggle to pass along them.

A game helps to sharpen our senses in this area. Look at the branches that birds choose to perch on and note how they change as the wind picks up. Try to predict the bird's next move. Pigeons are happy on the thin branches of a birch in a breeze but will hop over to sturdier-limbed trees if it grows at all gusty.

We know that the trees that line rivers are wet pioneers. They have to contend with high levels of light, wind, *and* water. These species, including the alder and willow families, tend to have thin, flexible branches: That's the only way to cope with the forces of wind and water.

At the other end of the spectrum, some trees hold a little thickness near the ends of their branches. They are happiest in established woods, with lots of fellow trees that offer good shelter. You may have spotted this effect in long-dead oak branches poking out of a green canopy, which is only possible when the branches hold some thickness.

EMPTY TREES

When we look at a tree from the outside in summer, it tricks us into believing it is full of leaves. But if you stand under it, near its trunk, and look upward, you'll soon see that the tree is mostly hollow and largely leafless. There are branches that reach out from the trunk and hold no leaves. Nearer the edge, lots of shorter branches hold all the leaves. As you will have guessed, this is all about light. There is very little near the trunk so no point in going to the expense of growing leaves there.

Branches perform a dual role: They have to reach out from the trunk toward the light and hold the leaves that will harvest it, but these are different tasks. This is why many tree species have two different branch types: long and short. The long branches stretch away from the trunk and act as a scaffold for the short branches, which hold the leaves. It's an effect that greatly improves those tree-climbing days—we can spend happy hours on the larger branches near the trunk, with lots of space and no twigs or leaves getting in our way.

Stand under a spruce or other conifer and you can look up at a hollow inner cone that almost perfectly reflects the shape of the whole tree. There are bare branches in this hollow zone, but no needles.

Once you have noticed how the inner parts of individual trees are bare of leaves, it is time to look for this effect on a larger scale. Seek out a dense copse full of trees in leaf—broadleaf in summer or conifer at other times of the year. Ideally the trees will be growing close together in a wood small enough that you can walk across it in less than five minutes.

Look at the pattern of the smaller branches and leaves on the trees at the edge of the wood. Now walk into the center and compare the pattern there. You will see that the trees at the edge of the wood have lots of smaller, leaf-bearing branches on the side of the tree that faces out of the wood, the outside edge. They will also have lots of leaves at the top of the canopy, but very few small branches or leaves on the side of the trees facing the interior. When you look at the trees in the center of the wood, you find that there are few branches or leaves on any of the sides of the trees, but a good covering near the top. These effects are logical: Light can reach one side and the tops of the trees at the edge of the wood, but not the sides of the trees at its heart. This pattern grows more elegant and intriguing when we step out of the trees and look at the combined effect: The wood has acted as a single tree does in the open. Small branches and leaves cover all sides of the wood and its top, but there are very few near the center. When growing close to each other, the patterns of leaf cover and the smaller branches of many trees combine to create a single effect in the canopy.

When it comes to the canopy, a dense wood is a tree and the center is empty.

This is closely related to another pattern called "the escape from the woods" effect that we will meet later in this chapter.

UP, DOWN, AND UP AGAIN

Do branches reach upward, point downward, or lie flat? Surely, after three hundred million years of evolution, there is some consensus about which is best. Not quite: It's a case of horses for courses. Branches that angle upward are more likely to reach light than if they point downward, but they're also much more vulnerable to heavy snow. There is a lot of variety across the species, but fortunately some patterns apply to most trees.

We can think of branches as the tree casting a net to catch light. If it cast the same-width net at all levels, the tree would look like a cylinder and only the top branches would get lots of light. It makes sense for the net to widen the lower down a tree we look. One of the ways in which the tree achieves this is by changing the angle at which the branches grow. In youth, they grow up, and as they mature, they grow out. This creates familiar patterns that are easy to spot.

The youngest branches tend to be the most upward-pointing. They sag a bit as they get older and gradually droop. The youngest branches are found at the top of the tree and the oldest at the bottom. This means that whichever species you are looking at, the branches nearest the top of the tree are most likely to point to the sky and the lowest to point to the ground. Horizontal branches may grow in between.

Once the lower, older branches have grown far enough out that they reach the light at the edge of the tree's canopy, there is no longer any need for them to keep going. They turn upward once more and relive their youth. Notice how many long, low branches turn up toward the sky near their ends.

Branches of a Spruce

These trends can be seen in most trees. It is more marked in conifers with low branches, and subtle in many broadleaves. The beeches and oaks near me barely drop below the horizontal, even in the oldest branches, and the upturn at the end is weak. However, the spruce branches point 45 degrees upward near the top, are flat halfway down, and point 45 degrees down near the ground, which they nearly touch. But the ends of those lower spruce branches rebel and reach for the sky.

Larches have long, gentle upward curves at the ends of their branches, fingers that curl up and beckon us closer. And the ashes I pass have such a distinctive upward curve at their extremities that they stand out from a distance, especially in winter.

When we look at a cat, a dog, a horse, a frog, a spider, or any other animal, we see the number of legs we expect (unless it's a most unfortunate creature). The majority of animals are genetically predestined to have a fixed number of legs. That is not true for tree branches: A different process is at work. This is why many misunderstand the branches they see.

When we look at the branches on a tree of any age, it's easy to imagine that they are the branches the tree was born to grow. But if the tree had started life somewhere else, we'd see different branches and a very different pattern. We wouldn't just see the same branches that have grown differently: We'd see branches that wouldn't exist if the tree wasn't in that spot. Each tree grows branches in response to the world it finds, and this is part of a tree's genius. In every mature tree there are hundreds of branches that never started life but might have done, and hundreds that started life and are long gone.

Let's explore this concept with a business analogy. Imagine that a successful US company has thirty branches in the US and wants to open five branches in the UK. It will be a costly and risky investment, so a lot of thought, analysis, and planning will go into the exact siting of those branches. Should they open one in Glasgow or not? If the company gets it wrong, it will be an expensive mistake. The chief executive will have to make the final call and they will have to use all their experience, wisdom, and data to get it right.

A tree doesn't have a chief executive, research, or data, but it somehow grows with a brilliant strategy. How does it manage that? Simple: It tries everything and lets most things fail. Trees have a big advantage: Opening a branch is easy and cheap.

If a tree was faced with the decision of opening branches of the US company in the UK, it would say: "Let's open a hundred very small branches all over the UK, let the successful ones grow and the failing ones fail." Ten years later there are only branches in the profitable places. And there would be one or two in quite strange places, locations that a decision-making executive might never have thought to try. For trees, the genius is not in the choice of place but in throwing the net very wide and seeing what works. They try growing a branch almost everywhere, but ruthlessly shut things down if they don't prosper. This process is called self-pruning and trees are always at it. The branches we see are there only because they haven't yet been terminated.

We can see this effect in almost all mature trees. If you plant a pine and come back about ten years later, you would find a tree that isn't very tall yet, maybe only 10 feet (3 m), and its branches still almost touch the ground. It would be impossible to walk under it. Come back in another ten years and the tree has grown, but with few branches near the ground. This is the moment where we can fall into the trap of thinking the branches have been lifted as the trunk gets taller, but we know that they don't. New branches form higher up and the lowest branches, the ones in the shade of the new higher branches, have been "self-pruned." The tree has lopped off its own limbs.

SOUTHERN EYES

Late last summer I was exploring a nature reserve called Snitterfield Bushes near Stratford-upon-Avon in Shakespeare country, in the center of England. I smiled on spotting some valerian flowers, a sign that there are rocks or concrete in the soil—the area has a wild feel, but that's recent: It stands on an old Second World War airfield.

After a long, fruitful day of investigating a fascinating pocket of nature, I was looking for a spot to spend the night. The trees, flowers, and mosses confirmed what I had started to suspect: The ground was too damp for a comfortable night. My shoulders sagged a touch as I conceded to myself that I would need to move on to a drier spot. This was a little disappointing near the end of a long day outdoors, so I sat on an old stump for a snack to summon the energy for a push up into the nearby hills. While I was eating some dried fruit, I stared, only partially in focus, at a wall of mixed trees in front of me. After about a quarter of an hour of not thinking of or looking for very much, I noticed the trees staring back at me.

After the energy from my snack had reached my brain, I realized what was going on. I suddenly appreciated what I was looking at and it was something I'd never taken the time to think about before.

And then a minor observation blossomed into a small epiphany. A new natural navigation clue seized me. Seconds later I had more energy than I needed. I could have run in circles with excitement. Instead, I paced frenetically around every tree in the area.

Trees have eyes on their southern side. Let me explain.

A branch that doesn't help leaves to harvest light serves no purpose and will be shed by the tree. Shading is the most common reason that a tree will terminate a branch. It is ironic that trees grow branches that shade their lower branches, which then serve no purpose. A tree can't move so this is the only way in which it can adapt to growing taller in an ever-changing canopy. Trees lose twigs and small branches regularly and will occasionally shed a larger branch, too.

When a larger branch is no longer productive, the tree gradually shuts it down and uses resins or gums to form a seal at the junction with the trunk. This is important because any opening at the trunk invites pathogens to enter and potentially kill the whole tree—branches are a highway to the center of the trunk. Once the joint is securely sealed, the branch is cut off from water and nutrients and dies very soon after. Then bark falls away from the dead branch—I'm sure you've spotted these bare, bark-free dead branches on trees. Slowly it falls prey to fungi and weakens until it snaps and drops, leaving a stump, which in turn rots and withers away.

Look at the bark on the trunk of a mature tree and you will soon spot the places where the dead branches once grew; they look slightly different in each species, but they resemble "eyes" in many. Some have a curved line over them resembling an eyebrow. The eyes are easiest to spot in smooth-barked trees in an area that is open to the south, but every tree has these marks.

Trees grow more branches on their southern side, as this is the brightest. As they get taller, they inevitably shed lots of branches on their southern side, too. This leaves a series of "eyes" looking at us from the south side of the tree. I now can't help but see them gazing back at me, especially from smooth bark, and you'll soon find the same.

It shocks me how many times southern eyes must have peered at me without my realizing it. It is all the more remarkable because we have evolved to notice eyes looking at us. To be fair, they are quite well camouflaged, until we know to look for them. Then they peer at us and mock our earlier myopia.

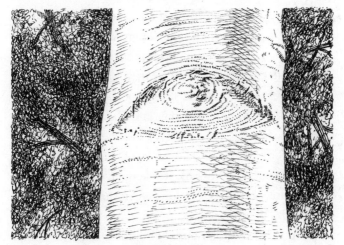

The Southern Eye

DEFENDER BRANCHES

Nature laughs at a hard rule. We know that trees are in the light-harvesting game and that branches that fail in this will soon die, but there are always small surprises to be found. Very low on tall trees, deep in the shade below a higher canopy, we will sometimes see small branches poking out of tall trees, often little above head height. What's the logic? There must be a reason.

Nature may not like hard rules, but evolution detests any waste of energy—plants don't use resources without good reason.

The low branches we sometimes see in shady areas are "defender branches." They are not there to harvest light for the tree, but rather to snuff out competition that may be trying its luck in the shade. There is no such thing as total shade in woodland—even in the densest rain forests there is still enough light to see where we're going. Woods aren't caves: Perfect shade would mean we'd need flashlights in the middle of the day and I've never been in a forest like that.

Defender branches steal some of the little light that reaches through the canopy. Remember, the slow strategy of shade-tolerant trees is to start life in shade and to reach sunlight by fighting slowly up to the top. If a competitor seedling was finding just enough light to start life under the high canopy, it soon gives up under defender branches.

Defender branches don't look the same as the main branches—you're unlikely to confuse them. There is a massive gap, a long bare trunk, between the two. High above our heads, thick branches lead up and out to the canopy layer. Then, near head height, one or two small branches may poke out. Defender branches tend to be horizontal: They don't reach upward because they don't care about the sky above them. They exist to hold an oppressive parasol over the already dim ground just beneath them.

PLAN B

Trees have dormant buds under their bark, sleeping tips ready to start life as a new branch. They can be found in many places on the tree, but are especially common near the base of the trunk, where it flares out and melds with the roots. The buds lie there under the bark biding their time, doing very little in normal times. (If bark peels away from the base of a tree, it is worth looking for these dormant buds: They are like pimples on the exposed bare wood.)

If a tree is struggling with its health, its hormones change, and a new message is sent to these normally shy buds. They spring into action from under the bark. They emerge as vigorous small green branches known as "epicormic sprouts." If you see an explosion of smaller branches bursting out from a tree's trunk or larger branches, you are looking at epicormic sprouts. They are a sign that the tree could be struggling, due to disease, damage, drought, fire, old age, or a combination of stressful events. Have a look at the top of the crown and you may see that it is not in robust health.

After a few years, a lot of the young sprouts die off, leaving only one or two branches thrusting upward from the side of the trunk. In woodland, these branches are often skinny and grow straight up toward the canopy, as that is where the only light will be found. (The steep upward angle makes these branches look different from the defender branches that reach outward to cast shade.)

If you see a thin, upright branch growing from low down on a tall tree and clinging more tightly to the trunk than you'd expect, it was once an epicormic sprout. Have a look for disease or damage higher up or near it and you may spot the reason it started life. It was one of these branches that forced me to the ground at the start of this chapter. There's a reason for the shape and location of every branch we see, but it's easier to spot when we keep our eyes open!

In a few species, like some lindens, those buds can't be held back and spring into life even in healthy trees. But in many trees the sprouts are a sign that the tree is in trouble: Epicormic branches are a tree's Plan B. As we have seen, trees aim to grow strongly from the top, then delegate authority to the lower branches a little over time: That is Plan A. But one of the trees' survival strategies is not to stick rigidly to any plan if it clearly isn't working. If the main high crown is struggling and isn't providing the energy the tree needs, it pulls the ripcord and throws out a hundred branches from the bottom. "It's just what they'd be least expecting!"

The sprouts near the base start as weedy green shoots, but they mean business. They are real branches, albeit immature, and will grow as much as the situation allows. Most die away, but some may survive to become sturdy branches, then alternative trunks.

It can be hard to work out the exact early history of a tree with many smaller trunks that meet near the ground, but there is a strong possibility that a healthy tree with one trunk resorted to Plan B many decades ago, and what we are looking at are the epicormic shoots that survived to become trunks.

Back in the Stone Age, humans learned how to work with this regenerative trick in trees—how to make the most of the Plan B sprouting. Coppicing is the practice of regularly harvesting the timber of young trees, like hazel, by cutting the stems near to the ground.

A similar traditional practice, "pollarding," uses the same process, but the tree is cut a little higher, at about head height. This yields a harvest of young timber but also protects the tree from grazing animals—all trees are vulnerable to animals when very young.

Coppicing and pollarding sound brutal and, done carelessly, would be fatal to some older trees, but many young broadleaves respond to this apparent savagery by growing a new crop of multiple sprouts that soon develop into healthy stems. Far from killing the trees, it perpetuates youth and prolongs their life. The Tolpuddle martyrs tree that we met near the start of the chapter is being very carefully pollarded as a way to offer it a longer life. Experts believe it could grant the tree an extra two centuries.

The harvested wood had many traditional uses, including fencing, tracks, or firewood. These days, coppicing and pollarding continue, but more as a woodland conservation technique than for the crop. The old and

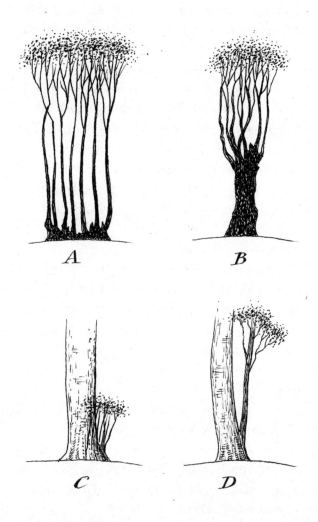

A—Coppice, B—Pollard, C—Epicormic Sprouts, D—Epicormic Branch

new use of these techniques may explain many of the interesting shapes we come across.

Epicormic sprouts are most common near the base of the trunk, but they can emerge from almost anywhere. A sycamore I see regularly is suffering from some ailment I haven't yet identified, possibly damage to the roots, and all of its main branches are in trouble. They stretch away from the trunk in a normal pattern, dividing into second or third orders, but none has any healthy leaves. Instead, there are perhaps a thousand thin vertical branches growing up from most of these branches and they each have their own leaves. It's an odd-looking tree, covered with greenery, desperate to cling to life, yet clearly in serious trouble. In winter it looks like a cross between a tree and a hedgehog.

You will also come across branches that emerge from the fork between two other branches. I see it in many species, perhaps birch most frequently. These in-between branches always look uncomfortable to me, like a sixth finger growing from the web between the thumb and index finger, but they're perfectly natural; there are plenty of dormant buds under the bark in these areas and a few break out.

A TRUNK-SHOOT COMPASS

I once spent a fun weekend with my wife, Sophie, in East Anglia, driving between the counties of Norfolk and Suffolk to watch our sons play sports matches. On the way to a hotel in Ipswich, we passed through a part of the country I had not explored and I begged to stop the car and walk into the trees. I had an uncanny feeling that a discovery was waiting to be made. I get this feeling sometimes. It isn't infallible, but it's hard to resist. Like most families, we know only a few things for certain: that I like going walkabout is one.

Setting off from the parked car, I walked along a disused railway into mixed woodland. Sophie decided to follow her own route: She is patient and generous, but not crazy. After a few minutes' taking photos of some towering cumulus clouds that pointed to the sun in a fabulously clear way, I decided to treat myself to an "invisible handrail" exercise. This is where we use a line in the landscape to wander freely, knowing that we can find our way back to it easily by following a different, as yet unknown, route.

I set off to the southwest and knew I could meander wherever whim took me, all the time secure in the knowledge that I could use the sun, trees, or clouds to head north again, meet the railway embankment, turn right, and find the car. There is something about the freedom that the invisible handrail gives: We know how to find our way back without resorting to maps or screens or having to follow any prescribed route or path. It opens parts of the mind that other plans can't. About ten minutes later I saw some oaks at the edge of a clearing and on them I discovered a new natural navigation compass.

Changes in light levels can trigger new shoots to grow from buds under the bark of the trunk in broadleaves. If an oak that has lived in shade for many years suddenly finds itself in light, it is likely that some new shoots will burst from under the bark of the trunk. Most sunlight comes from the south, so most of these shoots sprout from the south side of the trunk.

They don't look like normal or defender branches: They are much shorter and quite untidy, more like bushy stubble. I was standing by three trees that revealed perfect compasses, signs I had never noticed before. How many times had I struggled for direction near this powerful sign, blind to it? I dread to think, but now I see it all around me, and you can, too. Like someone fresh to spotting patterns in a river, it is easy to lament the downstream shapes that passed us by, but the joy is that more are always heading our way. And now we know how to see them.

After returning home I researched this effect. I'd never given it any thought before, but I rarely believe I'm the first to have noticed something. I may occasionally be the first to use something as a navigation aid, but that's very different from thinking it has hidden from human eyes for thousands of years.

Sure enough, I found an academic paper that referred to the effect as being common after thinning in woodland.

It is the same botanical process as the Plan B branches—epicormic sprouts—only this time triggered by new light, not ill-health. It even listed the species most likely to display this effect and, to my joy, oak was at the top of the list. Birch and ash were near the bottom. The name they gave it was "watersprouts," but it will always be the "trunk-shoot compass" to me.

ESCAPES, AVENUES, AND ISLANDS

Storms, disasters, disease, and people create many gaps in woodland. In any large new space, pioneers will start to fill the gap, but in smaller spaces, the branches of neighboring trees will grow longer to seize an opportunity.

Just like the apical bud at the top of the tree, the growing ends of branches respond to changes in light levels: This is how they wriggle into spaces but stop short of crowding each other in the canopy. Look up through the trees in a mature woodland and you'll spot a very thin line of sky that separates each tree's canopy, an effect called "crown shyness." (If the branches overgrow, they start to knock each other, which also stops growth.)

Five years ago, a forestry team removed one line of firs from a conifer plantation not far from my home. The branches on the sides of the trees bordering this new thin clearing look very different now: They have leapt into the space, closing out most of the new light.

You can see a similar effect at the edges of all woodland. Compare the branches on the inner and outer sides of trees at the edge of any wood and you'll see how different they are—struggling on the dark side, abnormally large and long on the light. I call this "the escape from the woods" effect: The branches seem to be making a bid for freedom from the woods. Sometimes the effect is so marked it appears that the branches on the woodland side of these trees are being killed by their neighbors, which in a sense they are.

The effect is doubled along any track or road that passes through a wood. These routes open up the canopy, forcing a clear line through the trees. The branches on either side leap into the space and try to fill it. I call this the "avenue effect," and it is exaggerated in these situations, because constant use means that pioneer trees can never fill the gap. In a wild context the branches would grow a bit and a few pioneers would burst up through the middle, stealing lots of the new light. Along well used routes, the higher branches on either side get a free pass and make the most of it.

The next thing that happens is that the branches start doing too well. They threaten the route they are growing over. At this point someone will deliberately or accidentally cut them back and branches will be knocked or lopped off. On busier roads by trees, you will hear the smacks, cracks, and rips as heavy goods vehicles or tractors do the job unwittingly. Almost every route through woodland shows the two avenue effects: exaggerated growth into the space, plus signs that branches have been cut back in some way.

It's important for natural navigators to know these effects, and studying them develops into a fun art form. In an open environment, larger branches on one side can be a strong indicator of more light and therefore that we're looking at the south side of the tree. But when we look at any grouping or line of trees, we have to be sensitive to the avenue effects. We'll see lots of large branches bursting out of the north side of woodlands, for example, because on the north side of woods the sky to the north is still brighter than the dark woods crowding them on their south side.

In some parts of the world, you will come across islands of trees on hilltops or in the center of large fields. There are some interesting historical reasons for this: They have escaped the axe for defensive, hunting or even taxation purposes. In some countries, including Germany, land with a few trees was once taxed more lightly than open fields.

We can make a couple of interesting observations when we find these micro-woods. The branches try to escape from all sides of the island, bursting out in all directions, and if we take the time, we will notice how they do it unevenly. They grow out determinedly on the brighter southern side, but the longest branches may well be on the downwind side, especially if it is a hilltop. If it is a compact wood, walk all the way round it and notice how branch lengths fluctuate and the character of the woodland edge changes every few paces.

I once walked across a stunning part of Dorset. I was rewarded for a long day's trek with a dramatic demonstration of island tree effects on the top of a local ridge, at Win Green Hill, in southwestern England. A few others had climbed to the top and were enjoying the fantastic views in all directions—we could look down into three counties and across the sea to the Isle of Wight—but my eyes settled on the island of beeches at the summit. I circumnavigated the trees and reveled in the patterns.

It is always thrilling to witness bold demonstrations of simple principles. The southern branches were strong and long, but the branches on the northeast side of the island tailed away, like streamers on a breeze. As I headed down the hill, I thought that if we filmed the edge of the island as we walked round them, then played the film back speeded up, we might see the branches breathing in and out. My mind soared and my heart was full of joy. Then I slipped on some damp chalk and nearly fell. That is how it goes.

You will see a particular type of avenue effect along riverbanks, where the branches of water-loving species throw themselves into the light over the water. The rivers form an avenue with a difference. Over rivers, the trees get a free pass: Humans leave these branches alone, and they can grow to extraordinary lengths. There are no chainsaws or rival trees, and the light over rivers is an untainted feast for any species that can stand the soaked soil at the bank.

During a recent visit to Snowdonia National Park in Wales, I made my way through the fresh damp November air of a Celtic rain forest. I was on my way to meet Alastair Hotchkiss, a conservation expert from the Woodland Trust. We spent a very enjoyable few hours meeting many of the rare and wonderful species that flourish in the mild moisture of the mountains near the west coast of Wales.

After a gentle climb over wet rocks, we stood close by a sublime waterfall that would have excited Wordsworth. (The tallest, grandest waterfalls attract most of the fanfare, but in my experience, small waterfalls that we can almost touch do more honest work on the soul.) A mist rose from the thunderous churn and floated past our faces into the trees where it nurtured the mosses and lichens. I looked down the course of the thin river and saw a charming example of the avenue effect. The branches of the trees wanted to come together and meet.

The branches of sessile oaks on each bank stretched over their wet, moss-draped roots and kept going until they were well above the noisy white water. It would be easy and lazy to presume that the branches met in the middle, but they didn't. The stream ran west to east, which meant the riverbanks were north and south-facing. The branches on the south-facing bank had made the most of the greater light from the sun and reached a little farther over the water.

OPPOSITE OR ALTERNATE

Most broadleaves favor one of two patterns of growth: Their branches either grow opposite another branch or they alternate. To see this for yourself, find some of the youngest branches you can get close to: Can you see another branch growing opposite each one? If yes, your tree has an opposite pattern; if no, it is probably alternate. (This will hold true for all branches of the same tree, but as the tree ages, it kills off many of its own branches, so we don't always see the clearest examples in its oldest parts.)

This pattern is repeated at many scales throughout a tree. If the leaves or buds are opposite each other, the small and larger branches will be, too. The same is true if the leaves alternate. Put another way, if you see two leaves growing opposite each other, zoom out, and you'll see lots of branches growing opposite each other, too.

Poplars, cherries, and oaks have leaves and branches that alternate. Maples and ashes have opposing leaves and branches.

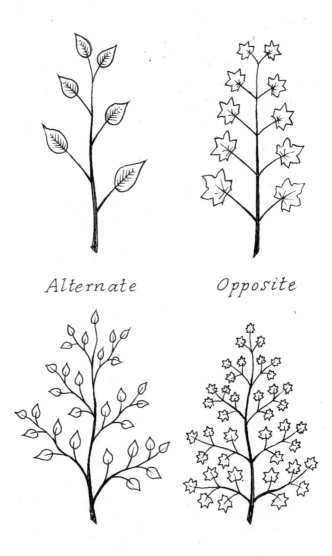

Alternate   *Opposite*

ZIGZAG

Each branch has a terminal bud that leads its growth, but they don't all behave in the same way. Some are more terminal than others. Some buds grow for a season, pause over winter, then grow again in spring. This leads to relatively straight branches. However, some terminal buds will grow only for one year. The most common reason for this is that the terminal bud forms a flower at the very end of the branch and that is the end of growth for that bud. The next spring, growth starts again, but from buds to the side of that flowering bud, which changes the angle of the branch at that point. The result is a tree with zigzag branches.

The formal name for straight branch growth is "monopodial" and for the zigzag effect, "sympodial." Beeches are monopodial and so are most conifers. Oaks are sympodial.

This is a good moment to pause and repeat the exercise we tried at the start of the chapter. Remember the tree I asked you to find and turn your back on? It may have been difficult to describe the branches in detail, but if you return to your tree, or find a new one, and look for the zigzag effect, you will start to notice patterns it was hard to see before. Your tree's branches will either look fairly straight and clean or crooked and chaotic. This exercise is easiest with bare trees in winter, but if you try it with a tree in leaf, go for an isolated specimen with a bright sky behind it.

Here is another quick exercise I'd like you to try with the same tree. Pick one of its major branches, follow it out from the trunk with your eyes, and see if you can predict exactly where it will end up. If this exercise is easy, you're probably looking at a monopodial tree; if it's fiendishly tricky, it's more likely a sympodial tree. Remember that sympodial tree branches change direction once a year: They can't hold a line. Following the branches of sympodial trees is like asking an over-eager stranger for directions in a town we don't know. It all blurs into "Left, right, right, then left, left again, right, next left . . ." In monopodial trees, it's more like, "Head away from the trunk and keep going until you find the light."

Monopodial trees still need lots of smaller branches, but they grow out from the sides of a main branch. Their terminal buds don't create a roadblock and change direction every year. When we look at monopodial trees in winter, we can usually see the major branches as dark lines that get thinner but run continuously all the way from the trunk to near the edge of the canopy. There is a wild cherry tree about 10 feet (3 m) from my cabin, and I can trace each of its branches to the very edge of the tree.

There's another way of gauging which you're confronting. Look halfway from the trunk to the edge of the canopy and see if you could, in theory, count the branches. In monopodial trees, it might not be easy, but you stand a chance; in sympodial trees, like the plane trees in many city parks . . . Good luck and maybe stop when you get to your first hundred! Whenever I come across a horse chestnut in winter, I imagine counting its branches and laugh at the ridiculousness of the idea.

Monopodial trees tend toward a more formal pyramid shape. Sympodial trees have a more rounded globular form. Sympodial branches are always alternate, never opposite.

Monopodial trees

Most conifers	Prunus family, including
Beech	cherry
Holly	Dogwood
Ash	

Sympodial trees

Plane	Linden
Oak	Sycamore
Maple	Willow
Birch	Malus family, including
Elm	apple

ORDERS

Lots of light means lots of little branches. This is a simple, beautiful pattern, but the explanation is less elegant. Bear with me.

A satellite image of a large river system, which flows from hills to the sea, will show one broad river near the coast and dozens of tiny streams up in the hills. And if we look at a diagram of the way blood flows from an artery into an organ like the liver, we find something similar, a large vessel at one end and dozens of smaller offshoots at the other. Each time a major vessel of any kind divides into smaller ones, we say there is another "order" of branching.

We find the same pattern in tree branches, which shouldn't surprise us because trees are the godparents of this way of describing the pattern. Almost all systems that subdivide invoke tree-branch analogies, from train lines, companies, and coral to family trees.

If a tree has only a few large branches and nothing else growing from it, we would say it has a single order of branches. But this never happens in living trees, because those large branches don't host leaves. (You might sometimes see it in a long-dead skeleton of a tree, where all the smaller branches have withered away.) Each time a smaller branch grows from a parent, we say that a tree has added another order of branches. If an even smaller branch grows from that second-order small branch, then the tree has three orders. How many orders do trees grow?

The basic rule is that a tree growing in full light needs to harvest light at lots of levels and from almost all directions. It will have lots of orders, as many as eight. That's a tiny branch with a parent branch, a grandparent, a great-grandparent—eight times! But if a tree has grown up in the shade of dense woodland, it may have as few as three orders. Imagine you are water flowing up the trunk and trying to reach a leaf. You may have to pass only three junctions to reach the leaf in a shady rain forest, but it may take another five turns before you reach a leaf in a sunlit pioneer tree.

The first-order branches, the ones that will ultimately grow to become the larger limbs attached to the trunk, have a primary aim: to get away from the trunk and toward the light. During this early stage, the last thing the tree wants is lots of orders: It would lead to a convoluted mess of branches more like a sponge than a tree. The tree has a clever trick that limits branch orders to a workable number. The growing bud at the tip of each first-order branch sends chemical messengers down the branch (just as the apical bud at the top of the tree does down the trunk). For the first year, these messengers—the same hormones called auxins—slam the brakes on and stop the branch developing second-order branches. After the first year, they relax and second-order branches develop.

That may sound complex and technical, but it leads to a clear pattern we can look for: "Lots of light means lots of little branches."

It's time for some mild exertion. I'd like you to pick up something heavy but manageable in one hand, maybe a large hardback book. Start a timer. Now lift it all the way over your head with one hand and hold it vertically above you until your arm feels tired. Bring the book down and stop the timer. Give your arm a good shake and rest for a few minutes. Now reset the timer, start it again and repeat the exercise, but this time instead of lifting the book above your head, hold it horizontally out to one side so that your arm is perfectly straight and the book is as far from your body as possible. End the exercise when your arm feels uncomfortable and stop the timer. Most people find the second exercise more demanding and shorter than the first. Trees do, too.

Trees have a problem with branches. The trees' trump card is their strong trunk that lifts them above the competition, but the trunk has no leaves so the tree needs branches to hold them. This creates more than a small challenge: Branches are built like the trunk, which has evolved to be strong and stable, growing near vertical. You can see the problem: If the trunk is ideally suited to a vertical form and branches grow out closer to horizontal, we have an architectural issue. Look at any big city and you'll find there is an area near the center with lots of high-rise buildings; some may be a hundred floors high. But nowhere on Earth will you find tall, thin buildings that have long side bits stretching out long distances. Branches can be thought of as small trunks forced to grow at angles that make engineering difficult.

Think back to our lifting exercise: When we hold the weight above us, our bones take much of the weight; the muscles do a little work and it is evenly spread. But when we hold it out horizontally, we soon feel that some muscles are having to work very hard, a lot harder than others. In the muscles near the top of the shoulder, in the spot where the arm joins the body, we can tell there is serious stress. The same is true in trees, because the physics is identical: Branches are just arms holding heavy weights horizontally from their trunk, which creates stresses and strains.

Dr. Claus Mattheck is a theoretical physicist turned tree expert. (One of his job titles was "Professor for Damage Science," which I hope, as a boy, he once dreamed of being.) Mattheck took a deep understanding of the causes and consequences of stresses in the world of physics and developed a new way of thinking about some of the shapes we see in trees. In a nutshell, trees don't like asymmetrical stresses and will grow wood in such places until it evens out the stresses.

If we held a weight out every day for years, our muscles would develop and grow to deal with it. One reason people go to the gym is to put on more "timber." Trees grow more wood wherever they sense a new strain: "Reaction wood" grows in response to a stress.

We felt the pain near our shoulder, and trees sense stress there, too. They grow extra wood at the point where the branch meets the trunk and this develops into an area known as the "branch collar." It is very tough wood. Historically it was used whenever strength was vital and has been found in Bronze Age axe handles.

Find any tree with a large, low, horizontal branch and look at the place where it meets the trunk: You'll notice it isn't a straight line. It flares out, growing wider at this point. Look more closely and you'll notice that the branch collar isn't symmetrical: The top and bottom are not identical.

Broadleaves and conifers both grow reaction wood, but they use different tactics. At this stage it's important that we understand two of the basic forces at work in all structures, including trees. There are two ways of holding something up against gravity: We either push it up from below or pull it from above.

Imagine you are moving a tall bookcase and it starts to fall toward you. Slightly panicked, you give it a big shove but then it starts to fall away from you, so you give it a gentler pull and it settles upright. Drama over and the bookcase doesn't fall, thanks to your quick reactions and use, first, of pushing (compression), then pulling (tension).

Conifers push branches up using "compression wood"; broadleaf trees pull them up using "tension wood." The cells in tension wood shorten, like tightening guy ropes on a tent. This changes many of the shapes we see in trees, including the branch collars. Conifers have a bigger bulge *under* the junction; deciduous trees have a larger growth *above* the junction. Wood is stronger in tension than compression but remarkably strong for its weight in both cases.

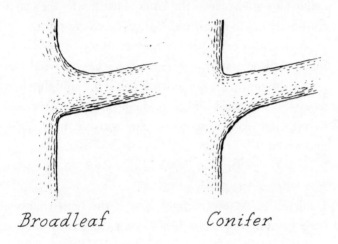

Broadleaf   Conifer

Why does the tree need to react to these situations? Wouldn't it be better to grow strong enough for all eventualities in the first place? A tree doesn't know beforehand how or where the greatest stresses will develop, and it would be inefficient to grow lots of extra wood that may never be needed. Unlike a person who goes to the gym often, then gives up, this is a one-way process: Trees don't grow wood one year and lose it the next. Once grown, wood is there to stay.

This is an important point: A tree cannot predict how long a branch will grow. As we have seen, light levels might cause a branch to self-prune when it is a young twig, or it might live until it is an old and mighty limb. It would be mad for a tree to grow the same size and shape branch collar for the former as the latter. That is why the collar constantly has to adapt and is always changing shape.

A tree also can't predict how great any other forces may be. Snow may weigh down branches, or wind can push them upward, even moving soil. If the land slips and a tree tilts, the trunk and every single branch will respond by growing reaction wood to cope with the new angles and stresses.

POINTING TO PIONEERS

Branches grow toward light, and if the light changes, their heading does, too. Once you have grown used to spotting how branches next to avenues and rivers show a strong preference to head away from darkness toward full light, you're ready to look for more subtle examples.

About a decade ago, I started to notice a strange but consistent bend from beech branches toward hazels in my local woods. It took me a few days to work out what was going on, but now that I understand it, I see the effect much more widely.

We expect branches to start to bend toward any new clearings in woods because of the sudden increase in light. But such gaps don't stay clear for long: Wildflowers, bushes, and pioneer trees soon race upward to steal the new light. They fill the gap. Before the gap is gobbled up, the established neighboring trees' branches bend toward it. The pioneer trees fill the gap, but the bend in the old branches doesn't disappear—once wood is formed, it is formed.

It's as if the old branches are pointing to the new pioneers, but their curve is a memory of a gap that has been grabbed by the younger trees.

WITCH'S BROOM

You will come across localized eruptions of tiny twigs bursting out of a branch, the "witch's broom": The dense collection of twigs can look a little like a traditional broom, although more often it's a lot messier. Witch's broom is a slightly chaotic defensive response by the tree. The precise cause varies from hormone problems to invading bacteria, fungi, or viruses, but the effect is a bundle of intertwined twigs, often with leaves caught in it. Enterprising animals sometimes make a nest in them.

I see them in my local woods, and they remind me that the tree's hormones normally do a great job of keeping order. Each witch's broom is a picture of the tangled mayhem that all trees would become if regulating hormones weren't telling each growing bud what to do and when.

It is miraculous that trees have come up with so many similar but different ways to grow hundreds of branches that fill a space efficiently, without total anarchy. Each species does this in its own way, but the golden rule is that branches grow at angles that allow them to head toward light and away from each other. There are always two parts to this: the genes and the environment. The genes tell the branches to grow away from the trunk and give them their rough pattern, but the light shapes the exact angles. That's why branches on the south side of a tree grow closer to horizontal, toward the sun, and those on the north side grow closer to vertical, toward the bright sky above them. This phenomenon is vital to natural navigation and I call it the "tick effect" (or "check mark effect")—seen from the side, the branches make a tick shape.

No system is perfect. Sometimes the branches get it "wrong" and grow too close to each other. Sometimes they even touch or crash into each other in slow motion. In an undisturbed calm environment, with no outside influences other than light, this is unlikely to happen. But animals, the wind, falling branches, disease, and a number of other challenges can set a branch on a collision course with another.

When the bark of one branch rests against the bark of another, at first nothing very exciting happens. Over time, though, the two branches will move in the wind, which causes friction. This rubs away the bark at the spot where they touch, the growing tissues below come into contact and join, or "meld," sharing resources and burdens. The formal name for this partnership is "inosculation."

When small branches join, it creates an interesting pattern, but it doesn't do the tree any harm or have major consequences. However, if major branches do this—or small melded branches grow into larger ones—it lights a long fuse on a structural time bomb.

We may imagine that two branches joined in this way are super-steady, and for a few years that may be true. The branches do support each other but, ironically, that is what causes the problem to develop: They don't grow the supporting wood they need to survive on their own. If you don't take the training wheels off a child's bike, they'll never learn to balance.

Eventually one of the two branches will weaken or drop and its partner won't have the strength to deal with the situation. Large melded branches are a problem waiting to happen, which is why tree surgeons operate on them. We will return to this issue when we look at forks in trees in the Bark Signs chapter.

5

Wind Footprints

Windthrow or Windsnap? • Harp Trees • Flagging •
Wedges, Wind Tunnels, and Lone Stragglers • Flexing •
The Awkward Tree • Mysterious Patterns •
Shadows, Eddies, and the Bulge

T HE WIND LEAVES FOOTPRINTS in the trees, some light, others more striking. A gentle breeze may bend the topmost twig over, but a gale will snap century-old roots out of the ground.

In this chapter we'll look at the different ways in which the wind changes trees, starting with the most violent and moving down the scale. At the end, we'll investigate some of the more mysterious effects you'll see.

WINDTHROW OR WINDSNAP?

On December 23, 2013, local people took shelter as a storm tore through Kent in South East England. After the worst had passed, Donna Bruxner-Randall discovered that the fierce winds had toppled a 39-foot (12 m) fir tree on her land. It had fallen at the edge of her property and now lay over the boundary of a neighboring farm. It had pivoted at its base and a large section of soil had lifted out of the ground with the roots. Tom Day, the neighboring farmer, wasn't too bothered by this and said he'd deal with it, but there was no hurry. The fir lay on its side for a month.

Then, on February 1, 2014, less than six weeks after the first storm, another hit the same area. When the winds died down, the locals headed out once more to inspect the damage, probably fearing the loss of more trees. Donna was in for a surprise. Her fir, the one that had come down in December, was now standing fully upright. The winds in the second storm had blown in the opposite direction to the first and pushed the tree up. Donna was as stunned as her neighbor.

"It's just so weird, as it's gone up perfectly. The farmer was gobsmacked and the second storm is the only explanation we have for it."

Almost a decade has passed since those storms and I was very curious to know how the tree had fared, so I got in touch and asked after it.

"It's still standing and actually looking very healthy!" Donna tells me. I detect a hint of pride in her miraculous tree.

This, you'll be aware, is not the normal course of events. Once a tree comes down, it usually stays down, but that doesn't mean it's dead.

Ferocious winds can bring a tree down, but they do it in one of two quite different ways. The most common is "windthrow," when the tree is uprooted but remains intact—the root ball twists out of the ground as the tree falls. It is especially common in spruces, but all trees will topple if the forces are strong enough. This was what happened to the Kent fir during the first storm. Windthrow is more likely after heavy rain that weakens the soil—the fir had been standing in sodden ground. If you see an uprooted tree, have a look to see if the roots broke, the soil lifted, or both. The roots frequently break on the side the tree falls toward—the downwind side. They buckle and snap as the trunk tips over.

Trees tend to fall in line with the wind and therefore form a trend over the whole area where the storm hit. Once you've identified the direction in which they're lying, storm-felled trees can give a strong sense of direction, even deep in dense woods. It's often the same as the prevailing wind direction—where the wind blows from most frequently—but not always: Storms can come in from any direction.

The second way in which the wind will bring down a tree is "windsnap": Here, the roots hold, but the trunk breaks. This rarely happens without a structural weakness in the trunk. Disease or earlier damage makes it more likely, and if it's a recent event, it's worth looking for signs of rot or fungi near the break. More often than not you'll be able to spot discoloration on the bark and in the wood near the fracture and sometimes fungi bursting out.

Windsnap is lethal and can kill mature trees. Windthrow isn't normally fatal: A tree stands a good chance of surviving, as long as a few of the major roots are still intact and anchored in the soil.

Conifers that survive will start growing again from their tip, the highest point before the tree came down. Broadleaves will try to make a new trunk out of the biggest surviving branch nearest the roots. This leads to interesting forms and shapes that can be deciphered many decades later, such as the harp tree.

HARP TREES

As we have seen, if a storm blows a tree down and it rotates from the root ball, it is likely to survive, as long as some roots remain intact. But now the tree needs a serious change of plan. All the branches on the underside that didn't die in the impact will soon die in the deep shade, leaving only branches on the upper side.

Sometimes epicormic sprouts on the upper side of the trunk are sparked into growth by the combination of stress and new light. This can lead to a striking pattern: It looks like a series of parallel smaller trees growing from an old horizontal trunk on the ground. It has a few nicknames, including "harp tree" and "phoenix tree," presumably as it appears to have risen from death.

A few years ago, I spent a day studying snow patterns on the lower slopes of the Cairngorms in the Scottish Highlands. It was a wonderful day, an intense one. I started with spotting big broad trends, like snow building up on one side of rocks and trees—after heavy snow, there are usually vertical strips stuck to one side of the trees, the side the snowstorm's winds came from. Once noted, it is a dependable trend, helpful in natural navigation.

As the day progressed, I shifted to looking for more subtle clues. By midafternoon I was searching for patterns at individual snowflake level—and mostly failing to find them. It required a level of concentration that was a little draining. Then, as the sun dipped behind a ridge, I took a break. I allowed my gaze to wander away from the minute and out to the broader landscape. And then I saw a sign shine from the trees.

The conifers along the ridge had a compass in their form: They all pointed the way so confidently that the bold simplicity of their message almost made me laugh. I don't yet understand the neuroscience of why a shift from narrow focus on detail to seeing grand conspicuous meaning in nature leads to euphoria, but it does. The name of the pattern I saw that cold afternoon is "flagging."

The wind can do a lot of harm to trees without killing them. Trees in exposed locations struggle, but certain branches suffer more than others. The highest branches

Flagging: In this picture the surviving branches are pointing away from the prevailing wind.

of the highest trees get the worst deal, and those on the side from which the prevailing wind comes often fail, leaving asymmetrical treetops with one side doing OK and one bare side. The surviving branches point away from the wind, which is why this effect is known as "flagging." In the midlatitudes of North America and Europe, the branches (flags) tend to point east. It's definitely worth looking for this when you're on hills or near the coast.

WEDGES, WIND TUNNELS, AND LONE STRAGGLERS

Trees react to the wind by growing shorter and stouter and showing a more pronounced taper in the trunk—it gets thinner higher up. This is one reason why the trees get taller as we walk into a wood: Those at the edge are more exposed to the wind and therefore shorter. It is also why trees on the side that the prevailing wind arrives from are the shortest of all, creating what I have nicknamed the "wedge effect." The wood slopes down in the direction that the prevailing winds have come from. (The wedge shape can look a little like the hood of a sports car, and we just need to remember that the car drives into the wind.) This is toward the southwest in the UK and many other northern temperate regions; in the midlatitudes of North America, it's often closer to west. Note that prevailing winds can be affected by local landscape features, but once you've determined your region's prevailing wind direction, it becomes another trend useful in natural navigation.

The "Wedge Effect": The trees on the windward side of any woodland will tend to grow less tall than the sheltered ones.

The wind sculpts woodlands, but it also changes the shape of individual trees, molding them into aerodynamic shapes. They tend to be shorter and more densely packed on the windward side, taller and more open on the downwind, leading to a form I call the "wind tunnel effect." When seen on a ridge, as a silhouette against the sky, you will notice that the side the wind hits is shorter, denser, and darker, but you can see the sky through the branches on the downwind side. There are also "lone straggler" branches on the downwind side, individual branches that stretch away from the main canopy and point downwind.

The "Wind Tunnel Effect": The prevailing winds have come from the left of the picture. Note the shape of the tree, but also that more light gets through on the downward side and the "lone straggler."

In extremis, the wind kills branches, as we saw when we looked at "flagging" earlier. But long before that, the wind has three more subtle effects on tree branches.

It bends the tops of trees in the direction that the prevailing wind blows. In the UK and much of the northern temperate zone, that is, from the southwest toward the northeast, and in the midlatitudes of North America, often closer to following a west-east pattern. This is one of the most effective techniques for natural navigators.

The wind shortens branches. Strong winds make trees grow less tall and the same process is at work on the branches. Trees in windy locations have shorter branches overall and they are shortest on the windward side.

The wind also bends branches, but the way it does this depends on the branch angles. The basic rule is that the wind will bend the branch toward the trunk on the windward side and away from the trunk on the downwind side. For example, if a windward branch is pointing upward, the wind will press it farther up, toward the trunk. On the opposite downwind side, the same wind would push the branch away from the trunk. Regardless of whether the branches are pointing upward or downward, those on the windward side end up closer to the trunk and those on the downwind are pushed away.

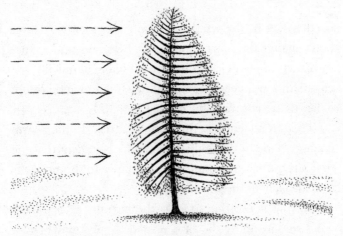

*Shorter branches bend toward the trunk on the windward side, with longer,
straighter branches on the downwind side.*

FLEXING

All of the effects we have looked at so far are the result of
powerful or long-term trends that leave lasting effects. But
changes also happen that last just a few seconds, before
the tree returns to its former shape.

Each tree family has its own character and reacts to
the wind in its own way. Some are stubborn and unyielding, others very flexible—a little like people in that sense.
On a windy day, take any opportunity you can to study
the different ways in which the trees meet this challenge.
There are characteristic responses to look for on every
scale, from individual leaves to the whole tree.

The leaves of many broadleaf pioneers, like poplars,
curl up into tight cylinders and let the wind slide by. Pine

needles flex in the wind, but many short conifer leaves don't change their form noticeably. It can be hard to spot the exact shapes of leaves in wind: It happens quickly, but color shifts can help—as we will see, the top and bottom of leaves are not the same color.

This trend carries through the tree. If the leaves change shape in the wind, you can be confident that the tree has branches and a trunk that taper to thinner ends that also flex in the wind.

Flexing is not the same as the harmful effects from wind we saw above, but it is related. A gust may bend a tree over for a few seconds and then, as soon as the wind dies down, it bounces back. However, if most of the strong winds come from the same direction, over time we will see wedges, wind tunnels and lone stragglers start to form.

By chance, as I'm writing this, a strong wind is passing through as a storm passes safely to the west. A spruce tree is bobbing quite happily, the branches bouncing up and down, but there are no major movements in the trunk or the leaves. The beeches are holding their shape low down but morphing at the higher levels: The outermost branches are squashed by each gust, then rebound immediately afterward. The leaves are twitching but only on the most exposed parts of the tree. The very top of the nearest beech's trunk is rocking gently.

The young birches are swaying with every gust, like yachts at sea in a gale, but unlike a ship's mast, their thin trunks are bending right over from a little over halfway up, then whipping back. The birch leaves are flapping so wildly it looks like they're disappearing and reappearing as they twist and flip back again very rapidly.

THE AWKWARD TREE

Each of the responses we have met in this chapter so far is straightforward when we think about it on its own, but nature likes to bundle some of them up, which can make reading them a little more challenging. Sometimes we have to work out whether we're looking at long-lasting imprints or the effects of the last few seconds.

With practice it grows easier to read the instant effects and the history of the wind in trees, but when you're starting out, I recommend beginning with the most violent impacts and moving down the scale to the instant (which is why I have structured the chapter in this way).

Look for trees that have come down in a storm and work out if it was windsnap or windthrow. Next, look for trees in exposed locations, on a hilltop or near the coast, if you get the chance. Ideally choose a calm day. Then you won't have to decipher which effects are thanks to today's winds and which have been developing for a decade. If you're in a more sheltered location, like a city, treat the tops of park trees like individuals for this exercise.

Study the shapes of any exposed trees you meet and try to get to know them from different angles. Note how their shape changes dramatically as your perspective shifts and how the tops have been bent over by the prevailing winds.

Once you've spent some time noticing long-term trends and are starting to find this comes naturally, you're ready to look for the "awkward tree." In the most exposed locations, only small trees thrive. On my local hilltops, hawthorns do well and are a match for any gale. But wherever you are there will be opportunities to find trees that live with strong winds. In cities, some spots are prone to gales: Look near tall buildings, rivers, at the end of long, straight streets, or where the wind funnels through gaps.

Whenever the wind blows from its common direction, the prevailing one, then it will reinforce the long-term trends. If the top of a tree is a little bent over on a calm day, the effect is exaggerated on any day when the wind blows from that same prevailing direction. But each month there will be days when the wind blows from a different and even, occasionally, the opposite direction. This creates an "awkward" look in exposed trees. It looks as if the tree has styled its hair, brushing it all over one way, then someone blows a hairdryer from the "wrong" direction. The tree goes from a smooth aerodynamic form, to something far less elegant. It looks uncomfortable and awkward.

The awkward tree is a sign that we should expect strange weather: Wind from the opposite direction to the typical tells us that an unusual weather system is passing through.

The Awkward Tree

The final challenge comes when we realize we will sometimes need to read the combined effects of the wind and the sun on the shape of a tree. If you think you can see that both have left their marks on a tree and it is proving challenging to know where to start, always give the wind precedence. It can leave impressions that override the sun, but it's very rarely the other way round.

MYSTERIOUS PATTERNS

The wind changes the trees, but the trees also change the wind. The strength and direction of the wind change dramatically near trees. If you see windswept patterns in the trees that you find confusing or counterintuitive, it will help to understand these changes. The best way to do this is to meet them personally.

When a wind encounters a tree or wood, it is forced to rise over it, which leads to an area of calm air on the side of the trees that the wind is coming from. After the same wind has passed the trees, there is another spot immediately downwind of the same trees that it can't reach, leading to another area of calm air. These still zones are known as "wind shadows," the places the wind cannot reach. Butterflies and other insects are busy in these spots in summer.

Whenever a wind passes any obstacle, the lowest part is slowed by friction, which causes the wind to start to roll over and spin. Imagine running at full speed and tripping: Your upper body continues forward at speed, but your legs are slowed down; the result is a tumble forward, sometimes all the way into a roll. This is what the wind does as it tries to pass over trees. The rolling wind is known as an "eddy." Eddies form every time the wind passes trees and they explain why the wind often appears gusty and to be blowing in all sorts of strange directions, especially just downwind of the trees.

If you step into a woodland on a windy day, it will be very noisy at the edge, but calm down as you get deeper in. We all expect this effect, but there is a more subtle one that few notice. In woodland, the winds are strongest over the tops of the trees—we hear the canopy rustling—and calm near the ground, but we can feel an interesting breeze between the two levels.

Winds in woodlands are stronger at head height than a little higher or lower down. If you are in a wood on a windy day, try to feel the breeze on your face, then reach down with your hand and you'll sense the wind stop near your knees. Then notice how any leaves or branches a little higher, about 10 feet (3 m) above head height, are not hit by the same wind as your face. This effect is called the "bulge," and it happens because the wind squeezes between the canopy and the ground. You can also feel the same effect on a single tree on a hot day, when you'll benefit from the shade and the air-conditioning sensation of the stronger breeze under the tree.

Once you are familiar with these effects, it is satisfying to seek them out. If we stand in a wind shadow on the downwind side of a wood, then move away from the trees, we can track down and explore the eddies. It's like a superpower: You can make the wind blow from almost any direction you want just by walking a few paces, then head back to the trees and turn it off.

The bulge travels more slowly than the wind over the top of the canopy, which means we can use the sounds and sights of the canopy to predict fluctuations in the bulge wind. On a windy day, try watching or listening to a wind tug at the tops of some tall woodland trees upwind of you, then count how many seconds it is before you feel the bulge breeze on your face.

Once you've taken the time to meet the tree-tripped winds, it can help make sense of many otherwise mysterious patterns in the trees. Any leaves at head height often look more ragged than those higher and lower in the wood, thanks to the bulge. Saplings and pioneers near a wood can look battered by the eddies from the main wood. In landscapes with a series of smaller woods there are many interesting wind patterns in the trees, as the eddies roll off one grove and play havoc with the next.

6

The Trunk

The Greeting Lean • Older and Fatter, Higher and Thinner •
Flaring and Tapering • Skinny into Wind •
Bell Bottoms and Fairy Houses • The Cushion •
Bulges and Ridges • Curves • Forks • Snapped

ASK ANYONE, YOUNG OR OLD, to draw a tree and you will likely find the trunk has little character. The picture will show a pair of lines that join the ground to a canopy of leaves. And yet, if we step outside, no two trunks look alike. There are curves, bulges, forks, and other patterns that offer up a world of rich diversity. In this chapter we will focus on the characteristics of tree trunks and what they mean. We will start with the broadest trends, the ones that affect the whole trunk, then narrow our focus.

THE GREETING LEAN

Many patterns in nature are invisible to most, but easy to spot, then hard to miss. The next time you walk along a broad track, road, or beside a river that cuts through trees, notice how the tree trunks lean toward you.

We saw how branches reach out over openings, like tracks and rivers, and the trunks play their part, too. If the opposite were true, the trunks would lean away from the wide ribbon of light through the forest and push all the branches further into the darker trees, which would be a terrible strategy.

The same trend can be seen at the edges of all woodland: The tree trunks lean slightly outward, but it is most satisfying to notice it as we walk through the trees. The effect can be especially powerful when we're walking uphill through deciduous trees in winter: a bright sky behind bare trees, offering silhouettes and a strong contrast.

I like to think of the trees leaning out to greet us. There isn't a cell in my brain that believes this to be true, but it means I remember to look for the pattern, which gives me a warm feeling each time I see it. Try it.

OLDER AND FATTER, HIGHER AND THINNER

Long before we're old enough to spell "circumference," we learn to read one tree sign instinctively: The bigger the trunk, the older the tree. The girth of a trunk is more dependable than height for gauging age: Height starts to lessen in ancient trees, but the trunk keeps getting fatter. Some of the very oldest trees are shorter and fatter than they were in their prime.

A million variables can influence the exact circumference of a tree, but a rough rule of thumb works. Trees growing in the open with a full and healthy crown gain an inch (2.5 cm) per year. So, a tree with an 8-foot (2.5 m) circumference is about a hundred years old in the open. Trees in woodland grow upward more determinedly, for light, so the same girth in a woodland tree would mean it was twice as old—two hundred years in this case. At the edge of woodland, we find trees that are half open, and the same measurement there would suggest something close to 150 years old.

These are rough estimates, but they hold for a wide variety of trees, and, rather wonderfully, apply to both broadleaf and coniferous trees. Purists will want to interject that there are exceptions, including the true giants, like redwoods, and trees grow faster when they're younger and slower as senility approaches, so errors creep in at the extremes.

The total amount of water flowing out of a river can't be any greater than all the water flowing into it from all the smaller sources. A similar truth is at work in trees: The total thickness of a tree's branches is roughly the same whatever its height. If we gathered all the twigs at the top of a tall tree into a perfectly tight bunch, it would be roughly the same size as the trunk. Leonardo da Vinci commented on this in *A Treatise on Painting*: "All the branches of a tree at every stage of its height when put together are equal in thickness to the trunk [below them]."

It is a simple idea, which helps explain why we see the trunk narrow significantly above major branch junctions: There is now less tree to supply with water and nutrients. But we can think of this in another way. As we saw earlier, trees grow extra wood to cope with extra weight or stress: The trunk has to be thicker below large branches.

FLARING AND TAPERING

The village of Weldon in Northamptonshire, near the center of England, was once surrounded by Rockingham Forest, an area used for royal hunting. Rockingham was not an easy wood to navigate: Many visitors took a wrong turn and lost their way. People have been getting lost in woods for millennia and will continue to do so, but Weldon contains an old, incredibly rare, and ingenious solution to this problem. It survives to this day.

The story goes that one traveler became very badly lost in Rockingham Forest and only managed to get their bearings and find their way out by spotting light coming from the church tower at Weldon. The hugely relieved and grateful traveler decided they wanted to save others from the same terror and paid for a more permanent fixture. On the top of St. Mary the Virgin Church in Weldon, a cupola was built to house candles or lanterns. It is the only working inland lighthouse in the UK.

One lighthouse can teach us how to read the shape of tree trunks. John Smeaton, the eighteenth-century

British instrument-maker and engineer, was responsible for designing a new lighthouse off the coast of Plymouth. Smeaton knew he had to design something that could withstand the harshest elements all day and night, through the seasons and for many years to come. It was a formidable task for any engineer, but a little less daunting for anyone who appreciated that Nature had already worked out how to create something tall that could survive storms. You need strong materials, a firm steady base, and the right shape. Smeaton based his design for the Eddystone Lighthouse on the shape of an oak's trunk. He knew that stone would do a better job against the relentless waves than wood, but the shape didn't need much improvement. It lasted over a century, from 1759 to 1877, and was replaced only because the rocks beneath it had started to erode and become unstable. The tower itself was doing just fine.*

Most tree trunks flare near their base, but some, including oaks, do so more markedly than others. The taller and older the tree, the bigger the flare. Tall trees have to contend with the strongest winds, but it is easy to underestimate how extreme this effect is. A tree that is a tiny bit taller than its neighbors gets no shelter at all and this is the part that deals with the very strongest winds. A little extra height leads to

---

*You can still visit Smeaton's Tower near Plymouth. It was dismantled, then rebuilt on land after the public agreed to pay for it. It has stood there since 1884, "in commemoration of one of the most successful, useful, and instructive works ever accomplished in civil engineering."

significantly greater forces against it and a much bigger flare at the base. With its prominent flaring, the 3,200-year-old, 247-foot (75 m) President tree, a giant sequoia in California's Sierra Nevada and the world's second-largest tree, is something of an ultimate example of this.

Every trunk tapers nearer the top of the tree, but the way they do this is a reflection of that species' character and this mirrors the trends in their branches. Pioneer species, like larches, birches, and alders, expect to grow up in an exposed windy location, their trunks thinning to whip-like stems. The slow and steady trees, like oaks and redwoods, taper more gradually, holding some thickness almost to the very top.

SKINNY INTO WIND

No trunk is a perfect and symmetrical cylinder. If we imagine slicing through a trunk at about head height, we might think we will see the cross-section as a circle, but it never quite is. Something always skews it, and, as always, there are three parts to the shape we find: genes, environment and time.

Some species are preprogrammed to rebel against perfect shapes. Yew trees do not do regularity—there isn't a yew trunk in the world with a perfectly circular cross-section. And many smaller species tear up the idea of a neat single circle by breaking out into multiple stems; hazel and alder are particularly keen on this. Multi-stem trees start as a tight bunch at ground level, but over time begin to fall apart, the stems growing away from each other.

Many healthy trees, including beeches and oaks, have a fairly regular section from about waist height to just below the lowest major branch. This section may appear at first to be round, but it is more likely to be oval.

The whole tree responds to wind and this includes the trunk. Most are "skinny into wind." Walk a few times around a large mature tree in an open area and you will soon see how the trunk appears to grow fatter, then thinner, then fatter again. (This is one of the reasons that foresters record tree size by circumference and not diameter.)

The trunk is skinniest when you look in line with the prevailing wind direction, and it appears fattest when you are looking across the direction of the prevailing wind.

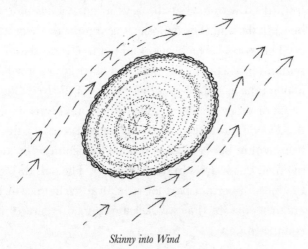

Skinny into Wind

Sometimes a tree trunk flows elegantly from the ground to the top, but we will often see swellings that break the smooth lines.

Let's start at the bottom and work our way up. We expect the base of the trunk to flare a little to give the tree stability, but some old trees have a base that looks outrageously fat. It's as if the bottom of the trunk has given up on being a tree and wants to become a giant bell. The names for this effect include "bell bottom," "basal bell," and "bottle-butt," but whatever name we give it, it's a sign of trouble inside.

Stop a mammal's heart, and you kill it. The same is true of many other inner organs, including the kidneys or liver. We have grown used to the idea that life is supported from deep within, that the keys to our vitality are well beneath the skin. The opposite is nearer the truth for trees.

The wood at the heart of an old tree is dead and if it remains protected by the bark and outer layers, it will last a long time as a stable but lifeless part of the tree. But if a crack or other weakness allows microorganisms into the dead wood, decay begins. Many ancient trees start to rot from within but can hold onto life for centuries by keeping their outer layers in working order. The outside is also the most important part for structural strength, which is another concept that we skeleton-framed creatures find counter-intuitive.

If there is trouble in the center of an old tree's base, the tree can survive by growing out and around the trouble. Ancient trees can profit from this process by reabsorbing some of the nutrients that end up back in the soil following the decay of their own interior. (Amazingly, they also grow roots within their trunks to feed off their own decay.)

Once more, trees solve a problem by growing more wood and the extra growth creates the bell shape at the base of these ancient trees. The crack or hole that allowed infection into the tree also widens over time. This means we will see magnificent old trees with holes, hollows, and other gaps deep in their lower trunks. This creates an effect you will have seen many times and possibly commented on, as it is quite charming. It's as if there is a fairy door into a small home at the base of the tree. Animals will often nest in these hollows, and kids love them, too, but these fantastical dwellings can be large enough for an adult to step into. I once huddled inside a fairy's house in a mighty elm, to shelter from cold rain. OK, I confess: The rain was my excuse; I did it because it made me happy.

THE CUSHION

On the bumpy road to making a living in the strange way I do, I once traveled to a meeting in Bedford Square, Bloomsbury. It's a part of London steeped in extraordinary literary history and overshadowed by grand Georgian buildings. For any aspiring writer, there can be few more exciting or intimidating places to head for a meeting.

I hate being late for anything, but this certainly seemed a bad time to change that habit. It was a meeting I felt sure would shape my life. I was eager, nervous, and arrived forty minutes early. Killing time by walking Bloomsbury in concentric circles, I found myself burning the last ten minutes circling the square itself, like a prisoner on a yard break.

At its heart, a formal, well-kept garden is fenced in by stout black metal railings. I would have liked to walk in the garden, sensing that the green space would calm me, but the gate was locked and I had no key. All I could do was pace around the metal spikes peering in at the greenery. I wanted to pass through the gaps and enjoy the other side. Then I saw that the trees had similar feelings: They, too, wanted to pass through the gaps. The base of a row of planes had swollen to engulf the base of the metal railings.

As a tree grows, the trunk fattens. If it meets something hard and unyielding, like a rock, brick, or iron railing, it will grow extra wood at that point: It forms a "cushion." The new wood forms a buttress at the contact point, although sometimes it engulfs the obstacle. Tree trunks can appear hungry to swallow anything in their way.

The meeting went well, I thought, and I waited for news. Such high hopes, a future mapped—albeit by the part of my mind that knows no cartography. When news came, it included the word "nice," which always means "no dice." It hadn't gone the way I'd hoped. It wasn't the pivotal moment I'd built it up to be, either. Such moments feel like they will be life-changing as we approach them, but hindsight allows us to see that the biggest moments creep up on us. Life went on, and after shedding feelings of rejection and dejection by walking long distances, I grew another layer of emotional wood to cushion me from the next hard obstacle.

BULGES AND RIDGES

Trees aren't aware of what the next question will be, but they know the answer. It's always "Grow more wood." Unlike many animals, which can regenerate their cells, trees can only add more.

Sometimes we will spot a bulge that wraps all the way around the trunk, well above the flared base. A swelling that encircles the whole trunk is a sign that the tree is trying to tackle an internal problem. The next thing to note is the character of the bulge: Does it rise and fall smoothly, like a gentle wave, or is it more of a sharp rise, like a step?

A smooth bulge is a sign of rot within that section of the trunk—the same problem as the bell bottom, only a bit higher. A sharper step is a sign that the wood fibers inside the tree have buckled, probably during a traumatic event like a storm. In each case the tree has sensed internal weakness and has grown a ring of new wood around it to shore things up, like a cast around a broken bone.

Whenever you spot a bulge in the trunk, it is worth looking for a sign of what caused the original problem. If the bulge is caused by rot inside, there must have been a way in for the attacking organisms, like an opening left by a branch that broke off without a proper seal. If there is no sign of an old branch, you may see signs of other trouble, like missing bark, possibly gnawed off by animals.

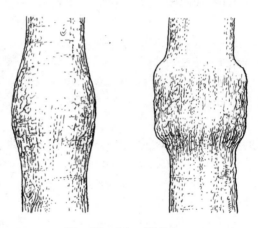

Wave (L) and Step (R) Bulges

Wood is one of nature's most amazing engineering inventions, but it has its limits. If the forces on it grow slowly and incrementally over years, trees add layers and cope with extraordinary levels of tension and compression. But wood cannot adapt instantly: If a storm, landslide, or other sudden shock hits a tree, the trunk may crack. A tree senses a major failure, like a crack, and, you guessed it, grows more wood to try to deal with the new weakness.

A major crack all the way through the trunk will lead to total failure sooner or later, but a crack on one side gives the tree a chance to recover. It grows wood around and over the problem leading to a raised ridge along the line of the crack. Sometimes this allows the tree to heal, but not always, and we can tell from the shape of the ridge how successful the recovery has been. A rounded, smooth ridge means the tree has healed; a sharp or pointed ridge means it has not. Horizontal cracks are due to tension in the trunk; vertical cracks form when there is compression.

If you try to break a small green branch, it won't snap cleanly or easily, but if you bend it violently one way and then another, you will see lots of vertical cracks then splits appear—an effect known as "greenstick fracture." You can often see light through a vertical slit in the stick long before it breaks in two. The compression causes a vertical crack, which widens into a larger fracture. The same crack forms in the trunk of an over-stressed tree, but before it can split completely, the tree grows wood around and over the crack, leading to the ribs we see on the bark.

The wind is the most likely cause of cracks and ribs, but there are other triggers. Freezing can create cracks in tree trunks, especially when one part of the tree is expanding or contracting faster than a neighboring part. Frost cracks are normally vertical.

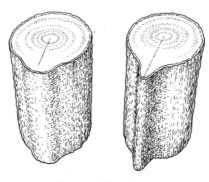

Smooth and Sharp Ridges

CURVES

On a cold, clear November night in Yorkshire, in northern England, I waved Rob and Dave off and wished them good luck. Their task was to find a farmhouse a couple of miles away using the stars, planets, moon, animals, and trees. They walked down the hill, and I felt a shiver of excitement as their silhouettes faded into the deepening blue of the early night sky.

I had spent the afternoon training the farming brothers for a TV program, and although there is a certain amount of fuss, fiddling and artifice in these situations, it is still always an exciting moment. I get to take off the armbands and nudge the fledglings out into the landscape, making encouraging noises as they try these techniques on their own for the first time.

I jumped into the Land Rover and took the longer road route to their destination to await their arrival in a few hours, all going well. They made it to the farm without the need for rescue and we were all grins and handshakes as our breath snaked up in the moonlight.

We debriefed, and they told me they had come off course a little in the woods, but had used the planets, stars, and tree shapes to work that out and get back on track. Seen from above, their track would have been a curve: They came steadily off course, then found their way back to the right heading. Tree trunks don't always follow a straight line: They sometimes come off course as they grow and have their own way of finding the right path again.

While I was waiting for Rob and Dave, I had kept warm by pacing the fields nearest to the farmhouse and this was when I saw the strongly curved tree. The silhouette of the Lawson's cypress stood against the bright background of the moonrise and its trunk was striking. It was only a little straighter than the average banana.

Trees sense gravity. The leading shoot at the top of the trunk will grow upward, in the opposite direction to the force of gravity, through a process called geotropism (also gravitropism). The clever thing is that the top of the tree is constantly sensing the force of gravity and adjusting its heading. This is important because a shoot can be knocked off course by many things and needs a way to get back on track. It can't turn round or start again: the past is permanently formed in wood. This means we can think of a tree trunk as the track the tree has followed: If it has come off course at any point we will see that in the shape of the trunk as a curve or sometimes a kink.

The curved cypress I saw that night was growing in the hills in the north of England. Chances were that heavy snowfall had weighed it down as a young tree, knocking it well off course. The long curve meant it had taken quite a few years to get back on track.

Many things will knock a tree off course, but snow and the land slipping are two of the most common. Look at the place where the curve is most dramatic: This will give you a good idea of when the event happened. If it is near the base, it was toward the start of the tree's life; higher up and it was later. If the change is very gradual and there isn't a single place where the curve seems sharpest, it may be that the tree has been steadily knocked off course, perhaps when the land beneath it slipped gently downhill.

A curve is not the same as a lean. There are times when a tree chooses not to grow perfectly vertically, which leads to a leaning tree, as we saw at the start of the chapter. The main reason is normally bright light from only one direction and this is common in three situations we have looked at: next to tracks or rivers and at the edge of woods. But it is also true on steep slopes.

Whenever we see leaning trees on steep ground, a gentle battle is going on between two types of growth. One is governed by gravity, geotropism, and the other by light, phototropism. The trunk's overriding aim is to grow upward, geotropism, but it might be nudged slightly off that heading by the extreme light asymmetry, phototropism. This is common on steep slopes as the light can only come from one side. Each species will prioritize each type of growth according to its preferred habitat. Riverside trees, like alders and willows, have evolved to prioritize light: They lean over the river and their trunks are rarely perfectly vertical. Most large trees and conifers will stay close to vertical unless physically forced off course. So, if you see a conifer with a curve or a lean, something more forceful than light has been at work.

If you look for these effects regularly, you will start to read the character of each tree in the way its trunk responds to light and gradients. Passing the edge of a mixed wood on a steep hill, for example, you may notice that some species lean right out and keep going; some lean out, then curve back to the vertical; and some grow vertically and appear to ignore these forces. This is actually quite an exciting moment because it means your powers of observation have leapt up. By the time we are noticing how each tree responds differently to the same influences, we are seeing things that few do.

Sloping ground is common, but it does lead to one more unusual and rarely spotted effect. On the slopes of hills, the taller trees tend to grow vertically upward, in the opposite direction to gravity. But the lower trees in the understory can actually harvest more light by growing out from the hillside, perpendicular to the land.

*Tall and Short Tree Angles
on a Steep Hillside*

FORKS

A mature and impressive beech stands a few paces from my cabin door. I have seen this tree many thousands of times and touched it many hundreds. On writing days I pass close enough to smell it, and in summer, I like to eat lunch in its shade.

One morning I set myself a challenge that I feared might be futile. I decided to spend a few minutes focusing on this one tree to see if I had overlooked anything. Were there any interesting features on this beech that I hadn't noticed before?

After looking from touching distance, I stepped back, and that was when I saw its shape properly for the first time. It shocked me. I was appalled by its *normality*. Behind my nearest neighbor, at the edge of the woods, there are two other beeches, but neither was nearly as neat. They were the same species and a similar size, but their shape was less classical somehow, less "ideal."

The two trees in the background were messier, lacking symmetry and beauty, but the nearest tree was perfectly sculpted for a beech. It appeared refined and elegant. I have noticed this pattern in a few other places in the area—the tree nearest a building looks neater than those a little farther away. The answer doesn't lie with gardeners or tree surgeons. It took a bit of pacing around the trees for me to crack what was going on. To explain it we need to spend some time with "forks."

Some trees live regular lives, meet with few calamities, and grow in the way their genes lead them. These lucky specimens form the archetypal shapes we see in typical drawings of trees. Tall trees prefer to have a single main trunk that runs from the base to the top without detours as this is the most stable shape. If the main trunk becomes two at any point, it has forked. Forks are an architectural weakness, which is why the tallest trees don't have them.

A fork in a tall tree is a sign that something serious happened to that tree in the past, typically decapitation. If a storm, animal, or human takes the top off a tree, new growth will start again near that point and often from more than one bud. If the new growth survives, the tree is likely to have two or more trunks. It is rare that a tall tree will support three equal-size trunks, but two-pronged forks are common.

As we know, the woody part of the trunk doesn't grow upward, so the height where the fork starts is a good clue to the timeframe of the incident that caused it. The general rule is that a fork that starts near the ground means grazing animals, like deer, are likely suspects. One that begins higher up could have been caused either by smaller animals, like squirrels or birds, or a violent minor catastrophe, like a storm.

A perfect tree does not exist, but anything that looks neat and idealized must have had a healthy top bud for most of its life, and this is reflected in the classical shape of a trunk that forms a single line from ground to summit. If the trunk is forked the tree has lost the apical bud, and this has residual effects beyond the fork. Remember, it is the apical bud that sends hormones down the trunk, stalling the growth of the lower branches and keeping the tree tall and thin, especially when it is young. This means that a fork suggests you look for the lower branches growing out more vigorously than they do from a single-trunk tree of the same species. A trunk with a low fork will lead to a wider, messier tree overall.

Back to my neighboring beeches. The nearest tree had had a healthy top bud all its life: hence its striking, classical normality. The messy beeches just beyond it had low forks, most likely caused by browsing deer, which have always been common in this part of the world. The deer like to browse a safe distance from buildings or other signs of humans—this is why there are more neat trees near civilization and more forked messy ones a little farther away. (Trees near buildings also attract more tree surgery, but that's a different story.)

Forks are points of weakness, but the earlier and lower they form, the more stable they tend to be. In the Bark Signs chapter, we will learn to look for the signs that show a fork may be close to breaking.

7

Stump Compasses
and Cake Slices

Cake Slices • Heartwood and Sapwood • Rings •
The Stump Compass • The Missing Stump • A Repellent Stump •
Spikes and Circles

THE ASH DIEBACK FUNGUS has claimed thousands of ashes on land near our home, and the trees that have survived remain extremely vulnerable. The public bodies that own this land worry about diseased trees falling, squashing people on public paths and creating a mess that then needs clearing up. Rather than lose sleep over this, they cut down ashes in their thousands.

I'm sure they've thought this through, and it's not for me to say whether or not it's the right policy, but it has robbed me of some of my favorite local trees and given me an opportunity to study countless examples of freshly cut tree stumps. I have learned to see many things in those stumps that I had never noticed before, and now I shall share them with you.

The first thing I look at is the bark around the base of the stump. If it is still tight, with no gap between it and the wood inside, it means life is left in the tree. Come back in a year and you may well spot the epicormic shoots we met earlier, sprouting up from around the base. If the bark has loosened and started to peel or fall away from the trunk, the game is up: The tree is dead.

CAKE SLICES

We inhale billions of mold spores each day, a lovely thought. They could grow into a fungus in our lungs that would soon suffocate and kill us, but they don't because our immune system kills them. We now take for granted that the air is swimming with viruses, bacteria, and fungal spores that would like to call us home. And we don't freak out about this every waking minute because we know we have defenses that work unbelievably well. However, this is still a novel concept.

For thousands of years, everybody could see what happens when pathogens succeed, but they couldn't see the bacteria, viruses, or fungi. Life was full of reminders that strange new life-forms can start almost anywhere, from mold on bread to someone dying from measles. It was as though these strange eruptions appeared spontaneously, which led the ancient Greek philosopher, Aristotle, to make a major mistake more than two thousand years ago.

Aristotle thought that life could start spontaneously from nonliving matter. He believed that many materials contained a substance, which he called "pneuma" or "vital heat," that could lead to new life beginning from these materials without any external influences. He pointed out that a clean, empty puddle, left for long enough, would soon be home to many living things. This theory of "spontaneous generation" explained how frogs appeared magically out of mud and mice out of moldy grain. It would also appear to make some sense of wood rotting and fungi sprouting out of it.

These days, we know that spontaneous generation is impossible and every new life on Earth has a parent of some kind, even if it is as basic as a virus. This awareness can help make sense of some of the patterns we will see in tree stumps.

As long as the theory held that wood decays spontaneously and with no help from external organisms, there was no reason to look for how trees defend themselves against pathogens. Perspectives shifted early in the twentieth century when the German forester, Robert Hartig, realized that wood decayed when infection set in and happened when a tree was invaded by fungi. It is obvious to us now, but this was revolutionary at the time.

The American biologist and tree specialist Alex Shigo built on this new insight and pioneered a view of how trees respond to infection. He noticed that when a fungus invaded part of the tree, the tree's response was to try to contain it. On detecting a pathogen, the tree reinforces cellular walls within the trunk to lock any infection into a chamber. Shigo called this process the "Compartmentalization of Decay in Trees" or CODIT.

There is a wall that stops the fungus traveling vertically up or down the trunk and one that stops it traveling toward the center. The one we will see most regularly with our own eyes is the "cake slice": The tree reinforces the radial walls, which travel from the center out to the edges of the trunk, like the spokes of a wheel. This contains any infection in a wedge or cake-slice shape in the trunk. Look at enough tree stumps or stacked timber in woods and you'll soon see a darker "cake slice." The slice is an infection locked in a wedge-shaped chamber.

Whenever we see the cake slice, we can admire the way the tree did its best to contain the problem. Unfortunately, we can see the wood because the tree has come down—a sign that it may have delayed the spread of the fungus but failed to stop it altogether.

The radial cells that contain the infection in cake slices also give the trunk and branches huge strength. This is why we see logs chopped into similar cake-slice shapes. (It is also the reason why green branches are hard to snap and we see the "greenstick fracture" we met earlier.)

We see the radial lines clearly when there is infection, but oaks are one of the only trees that show them in their wood without any infection.

Beneath the outer bark there is an inner bark, a thin layer of living cells that form an important tissue called "phloem." This layer transports the sugars made during photosynthesis and forms the vital energy network of the tree. It allows growth and works in the places of a tree that need energy but don't produce it, like the roots. The phloem wraps all the way around the tree but it is thin and close to the outside of the tree, making it vulnerable to any damage to the bark.

Under the phloem, the tree has a very thin layer of cells, too thin to see with the naked eye, called the cambium. This is the layer that makes new cells and drives growth, allowing the branches, trunk, and roots to grow thicker each year.

Inside the cambium we have most of the trunk. It is called the "xylem" and is made up of two parts, old

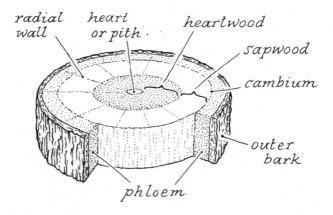

and new. Directly beneath the cambium there are young xylem cells, which are very much alive and busy transporting water and minerals up and down the tree.

Each year the tree adds a fresh new xylem layer outside that of the previous year. This is why we see rings and why the oldest rings are nearest the center. The xylem cells have a useful lifespan, but when enough layers have been added, the inner layers are no longer needed and die, at which point many trees fill them with protective gums or resins. The younger, outer, living layers of the xylem are called "sapwood" and the inner layers, "heartwood."

To the naked eye, the heartwood doesn't always look different, but in most species it is darker and, in some, it stands out dramatically, sometimes even being mistaken for disease. The very dark, dense wood known as ebony refers to the heartwood of some tropical trees; the sapwood is no darker in them than in any other tree. Spruce trees show very little contrast between heartwood and sapwood.

In some trees the heartwood breaks out of the rings, leading to more irregular patterns. External stresses, like drought, can alter the formation of heartwood, leading to striking shapes. I have seen stars, clouds, a chicken and even a panda in the heartwood of trees like beeches, birches, maples, and ashes.

Heartwood is denser, drier, harder, and heavier than sapwood so it is preferred for many practical uses. (Ebony is no longer favored for commercial use because it cannot be harvested sustainably in volume, but it remains a fascinating timber, so dense it sinks in water.) Some craftspeople use wood that combines both sapwood and heartwood to create beautiful effects, like a turned bowl from one piece of wood that naturally has both light and dark areas. The perfect longbow also contained sapwood and heartwood: The tensile forces at the joint added more zip to the arrow.

RINGS

Tree rings might help explain one of the most dramatic events in western history. In the late fourth century CE, the Roman Empire started to collapse with the help of marauding migrants from the east, including the Huns, a nomadic people with an infamous leader, Attila.

Dendrochronologists—tree-ring gurus—have found evidence of a great drought affecting China in the fourth century. There is a band of skinnier rings in the trees that grew on the Tibetan plateau during this period. The theory goes that decades of suddenly hot, dry weather forced the inhabitants to head west to seek moister, more fertile land, which led to the collapse of the Roman Empire and the arrival of the Dark Ages.

Most children still know that we can age a tree by counting the rings, I hope. The reason we can see the rings is obvious in hindsight, but few think of it when they look at them: There are two colors in each ring. If every year's growth had the same color, we would struggle to see the rings at all.

All the photographs in this book were taken by me—many very close to where I live in West Sussex, in South East England.

A sun-loving pine has shed its lower branches. Shade-tolerant beeches have kept theirs.

Defender branches. Lower branches are in leaf in spring before higher ones.

Trunk-shoot compass on an oak. We are looking west. Notice also the "tick effect" (or "check mark effect") in the branches on the tree to the left.

The island effect on Win Green Hill in southwestern England. We are looking north. It is darker on the left, windward side, and the branches reach farther on the right, downwind side. Notice also the "lone straggler" branch to the far right.

Trunks lean away from walls toward the light in London.

Dotty, our Jack Russell, inspects a harp, or phoenix, tree.

A fairy house in an old elm tree.

A bell bottom and a wave bulge in a struggling London plane tree.

Heartwood and sapwood and the "lonely hearts" effect.

A cake slice infection can be seen in the stump of a felled ash tree (ash dieback fungus).

The goats have created a "browse line" in a hawthorn in the Spanish mountains.

An oak that once had another tree growing to its left shows dramatic asymmetry, reaching for the southern light to its right—"the Repellent Stump."

Trampling over roots has killed the branches of a pine on sandy "common" land in Sussex, near the south coast of England.

The avenue effect in the branches of beech trees.

Alder trees lean over a river. The branches are held above the water level.

Each year the tree has a fast and slower stage of growing. In spring and early summer, it adds large cells with thin walls, which show as the wider, lighter part of the ring. Later, in the growing season, things slow down. The tree adds smaller, denser cells, which show as a thinner, darker part. The thin, dark growth acts as a divider, making it easy for us to see and count the wider, lighter part. Environmental conditions vary from year to year and this influences the width of each ring. A growing season that is kind to trees leads to fat tree rings; many imagine the fattest come after constant sunshine, but most trees grow best when it's moist and mild with enough light.

When we first start looking at the rings in the stumps of our local trees it seems miraculous that there are messages in them: They all look so similar. How do the dendro-wizards find any meaning in this odd language? They use a simple trick that we can copy: They don't look at the rings as one big group but for certain telltale lines. In every part of the world and during every era there are anomalous years, dates that mark very odd seasons. One of the early ones that helped pioneer this method in Europe was called the "Great Frost of 1709," a year that was so abnormally harsh it left its mark in a single ring on the trees of England, France, Germany, Sweden, and beyond. Wherever we are in the world, this is a time-travel technique we can try for ourselves.

Trees grow from the outside so the outermost ring is the final year of growth; the innermost ring is from its earliest youth. When you see a fresh stump with clean tree rings, make a mental note of which ring or pair of rings stands out most clearly to you. Now count from the outside in and subtract that from the year the tree came down: That is your local marker date.

Have a look for it in other stumps or large logs in the area. If a season was severe enough to leave its mark in the wood of one tree, it will have left its stamp in the others, too. It is fun to research what happened in that year: A bizarre summer, abnormally hot and dry, or a washout are likely. The seasons of 1975–76 and 1989–90 were especially harsh for trees in the UK and led to a double marker that I nicknamed the "twelve-year sandwich." There are records dating back over twelve thousand years in the oaks and pines of Europe. Unless your job title includes the word "dendro," I wouldn't recommend focusing your energy any earlier than the past century.

Each species has its own patterns, and the faster it grows, the wider the rings must be on average. Conifers grow faster than broadleaves so their rings tend to be wider (this is also why conifers are known as "softwoods"—the faster a tree grows, the less dense its wood on average). Don't bother looking for these patterns in the tropics; the trees can grow year-round, so the rings aren't worth searching for. Weather and climate are the broad forces that govern the width of each ring, but other factors are at work.

The general rule is simple: Stress reduces growth and makes rings skinnier. Stress comes in many forms and it's not always negative. In broadleaves the rings are skinnier in "mast years," when big-seed trees, like oaks and beeches, produce vast numbers of seeds. Parenthood can be stressful. We will look at mast years in more detail in The Hidden Seasons chapter.

THE STUMP COMPASS

I have found myself uttering a nonsensical sentence from time to time: "The middle of a tree is not in the middle." It's a slip of the tongue: Of course the middle of a tree is in the middle. What I mean is that the *heart* of a tree is not in the middle.

The heart is the oldest part of the tree. It's the bit we find if we follow the rings all the way in from the bark until we can't go any farther. (The formal name for the very center of the trunk is the "pith," but I will stick to "heart" here as it's more intuitive and memorable.) The heart is rarely perfectly central and nearly always slightly closer to one side of the trunk than the others. There are good and useful reasons why that is.

We have seen how trees grow reaction wood whenever they are stressed, adding compression wood in conifers and tension wood in broadleaves. Reaction-wood rings are wider than normal rings and their growth is always asymmetrical, which is the whole point: The tree is trying to offset a force that is pushing or pulling one way more than another. This means more reaction wood grows on one side of the heart than the other and explains why the heart is rarely in the very center of the trunk.

Trees in full light grow bigger and longer branches on their southern side as this is where most of the light comes from. The extra weight on one side creates stresses in the trunk and the tree grows more wood on one side to counter the imbalance. In broadleaves we should expect to find the heart nearer the southern side. In theory, the opposite is true for conifers with a heart nearer the northern side, but they tend to grow more evenly overall, so the solar effect is muted in conifers.

If light was the only thing to consider, things would be much simpler, but less interesting. The prevailing wind pushes trees more from one side than the others, which leads wood to grow to balance this force. The heart will be found nearer the windward side in conifers and the downwind side in broadleaves. (This is one of the reasons why the trunk appears thinner when viewed in line with the wind and fatter when looking across it.)

Land is never perfectly flat and the gradient has a huge impact on the location of a tree's heart. In broadleaves, the heart is nearer the downhill edge; in conifers, it is on the uphill side.

The final thing to look for is something I call the "lonely hearts effect." We know that a very young tree that loses its top bud near the ground will probably survive and sprout new shoots. Many years later this may lead to a fork and more than one trunk. These trees are always less stable than their single-trunked cousins, and foresters often cut them down. Notice how the hearts in these trees are nearer the center of the group, that is, nearer to the other trunks. Almost as if they miss each other.

All these factors play off each other, so we often see them mixed together. When you are new to them, I'd recommend looking for simple, dramatic examples. If you find a fresh stump in woodland on a steep slope, you're in business because sun and wind influence will be minor and the gradient effect will leap out.

Stumps lend us an X-ray machine and allow us to see many things that are invisible when the tree is in good health. It's madness not to make the most of it. I'm sure you'll feel a quiet joy rising in your sap the first time you notice how tension wood is a lighter color and compression wood, which has a higher lignin content, is darker. That is unless you're a woodworker, in which case you will dislike both as they distort when seasoned. Tension wood also forms a rough texture when machined.

Every species has its own grain and it's always good sport to try to identify a tree by the wood of its raw stump. Some give stronger hits than others: Cherry wood has a rich red hue and alder turns a vivid red soon after exposure to air.

We can also gather clues using our noses. Pine wood contains a resin with a pleasant but acerbic smell; yew stumps don't give off much scent. If you come across a yew stump, you might be tempted to try to age it by counting the rings. Get ready for a challenge, yew rings are one of the hardest to gauge.

There are also clues in the way a stump ages and, helpfully for our memory, it reflects how a tree lived. The wood and bark of fast-growing trees, like birch, cherry and ash, decay rapidly. Oaks hold tannins in their wood that delay decay, allowing them to age slowly and gracefully. The resins in the pine wood that give it a strong smell also hold off decay for longer than many other trees can.

Conifers evolved before broadleaves and have a simpler structure, which can be seen in the grain. Conifer stumps tend to rot from the outside inward, broadleaves from the inside outward. Cedars are the conifer exception: They rot from the inside.

THE MISSING STUMP

Occasionally you may spot a tree that looks like it's on short stilts, with roots that appear to lift the trunk off the ground. What strange magic could have caused this?

Healthy living wood has some natural resistance to infection, but when rot has taken hold of a stump, it breaks down the tissues and creates a friendly environment for new life. Seeds from other trees can use the nutrients in a decaying stump to fuel their own start to life, almost as though the stump has become a compost-filled flowerpot. More properly, they are called "nurse stumps" or, when the same thing happens on decaying fallen trunks, "nurse logs."

Over time the new tree puts out roots that spread over and around the decaying stump. Eventually the old stump rots away altogether leaving a new tree with a strange base and roots that arch over an empty space. Elves and fairies definitely make use of this architecture, but don't explain it to them: They prefer the mysterious.

A REPELLENT STUMP

While I was writing this chapter there was a night of heavy snow showers. My habit is to get up and out early at these times because I don't get many opportunities to study the rich variety of snow signs in the south of England.

After I'd had the joy of navigating across woodland using the strips of snow on the north side of the trees, the sun rose in the southeast. Rich pinks and oranges bounced back off a bank of cloud to the northwest.

By lunchtime the snow was in full thaw, and by teatime it had mostly vanished, except on the tops of the hills, where it had built up into small piles on the north side of each tree. As I neared home, I walked for several minutes without seeing any thick snow at all. Then I saw a thin layer of perfect icing on a broad ash stump. It stood out because there was no other snow anywhere near it. It was a beautiful little puzzle: Why was there thick snow in this one spot, but none elsewhere in all directions?

There were three parts to solving this riddle. The first is that the ground had warmed in the occasional sunlight of the day and the stump had made a fridge by lifting the snow off the warming ground. The air a few feet off the ground was cooler than the ground: I could feel it in my ungloved fingers.

The stump had also acted as an insulator, stopping the warmth in the ground reaching the snow. The final part of the solution is the most interesting for tree-readers. A large stump means the local skyscape has changed dramatically. There was snow on this one particular stump because there was no tree cover above it and the snow could fall freely to the ground. But this is something with much broader impact than snow.

Whenever we see a large stump, we can look for the way that the missing tree has changed the landscape. If there are neighboring trees, look for the "footprint" of your missing tree in their shape. A tall solitary oak I know well looks bizarre: It is bent over toward the south and has no branches of any size on its northern side. It is tempting to think that the light has sculpted this strange form all on its own, but it looks too extreme and odd for that. The answer lies in the mighty stump about 26 feet (8 m) to the north of the oak. Until recently, that oak was hidden behind a mighty ash that had been felled. The stump is all that is left of a large tree that had shaded its neighbor for many decades, leading to its lopsided and peculiar form.

If you look for them, you will soon spot branches that curve away from a large stump. The effect is the simple result of the shade that was cast by a tree that is no longer there, but I like to think of it a different way. It's almost as if living trees don't like large stumps, their branches finding them repellent.

SPIKES AND CIRCLES

The next time you see a fresh-looking tree stump, draw closer and take a good look. If it is gnarly and very rough, the tree trunk may have snapped during a storm. The majority will reveal a level surface because foresters or tree surgeons have cut them down deliberately. If you look carefully at the smooth stumps, you will often see the "spike."

When foresters fell a tree, they saw through most of the trunk, which leaves the flat, level parts that we expect to see. There will be a few lines, grooves, and notches where the saw pulled back, but most of the surface is a fairly clean cut. During the final seconds before the tree comes down, the feller will step away to a safe distance. The tree is now held up only by a weak, very thin section of trunk that hasn't been sawn. It isn't strong enough to keep the tree vertical and the tree starts to topple. As it falls, this skinny un-sawn section gives way and snaps—this is the violent cracking sound you'll hear if you're nearby when the tree comes down. It leaves a narrow, jagged short spike of wood jutting up from the stump. I enjoy looking for these spikes and then, on spotting one, imagining the violent cracking, tearing sound of the tree's last seconds connected to its roots.

Many trees have climbers, like ivy, growing up the trunk. When foresters fell these trees, they rarely remove the climbers as a chainsaw makes light work of them. We can see the stems as smaller circles of wood nestled up next to the main stump.

Sometimes a tree will grow wood partially or all the way around the stem of a climber—remember the "cushion" from the last chapter. This can make interesting patterns at the edge of the trunk of a living tree, but it also creates intriguing ones within the stumps of dead trees.

An ash tree I passed almost every day for many years had fully enveloped several ivy stems, but this was not something I could see when the tree was upright. Soon after it came down, I saw the stems as small circles inside the edge of the larger stump. It was a little like those close-up images of Jupiter's surface, where the large planet engulfs the smaller circles.

<u>8</u>

The Roots

THE DEATH AND DESIRE PATH

I was early for an appointment with Kevin Martin, manager of arboriculture at Kew Gardens in South West London. The Royal Botanic Gardens at Kew has superstar status in the plant world. It is a globally renowned center for botanical research with at least fifty thousand plants under its care, as well as being a UNESCO World Heritage Site. The team at Kew knows a thing or two about trees.

Kevin greeted me at the gate, and we didn't stop talking about trees or investigating them for the following two hours. It was a very happy time. Kevin didn't need to discuss his credentials: His tree-guru status was clear enough in his job title, but my CV is weirder, so I explained to him that I had been researching clues in trees, especially those related to natural navigation, for more than twenty years. We enjoyed comparing our thoughts about some of the landmark research and the individuals behind it.

For many years I have known that there is a strong relationship between the health of a tree's roots and its canopy directly above the roots. If you damage the root system on one side of a tree, the canopy directly above it will suffer most, struggling to put out leaves or dying back altogether. I need to understand the shape of tree canopies for natural navigation, so this fact has always been important. Is the canopy on one side struggling because of a lack of sunlight, the wind or because heavy boots have trampled the ground? For decades the relationship between roots and canopies had remained a dry fact to me: sometimes useful for solving puzzles, but not the sort of thing I'd actively look for. Kevin was about to change that by talking about a pattern in the ground known as a "desire path."*

---

*All my life I've been fascinated by paths. Paths are to a navigator what musical scores are to a conductor. I'm so besotted by these lines in the land that I've even come up with a name for a type of path. The "smile path" is my name for a curved path that goes round an obstacle, like a fallen tree or a large puddle. Smile paths are never shortcuts: They always follow a longer route, which is what gives them their curved "smile." They are everywhere, you've probably walked on a smile path in the past day or two, but they are little known or commented upon. I used to call them "bananas," but "smile path" is a prettier and better name, and the one endorsed by the Royal Institute of Navigation in 2020.

Desire paths form when pedestrians follow a popular shortcut. If a landscaper creates a path by laying stones through a lawn for people to follow, but walkers cut across the grass to save time, they will wear a line in the grass, which will show as a desire path. The landscaper wanted people to follow a certain line, but the new path reveals the route they truly *desired*.

Early in our walk together, Kevin led me to Kew's largest tree by volume.

"You see that label?" he said, pointing to a small rectangle of black plastic nailed to the tree's bark. I took a step toward it and read the white lettering:

CHESTNUT-LEAVED OAK.
Quercus castaneifolia
Caucasus, Iran

"Yes," I said, not sure what he was hoping I'd glean from it. "Visitors love this tree and they want to know more about it.

They used to head over to read that sign and they would all follow the same direct route. Thousands of feet marched over the same line to that little sign on the tree. There used to be a well-worn path in the ground—you can still see it."

I looked at the ground and saw the ghost of a desire path.

"We had to rope it off and move that plastic label. The footfall was killing the tree. You see the large branch that has failed." Kevin pointed to a spot over our heads where the rupture trauma of a fallen, large branch was clear. He explained that the constant compaction of all those feet walking over the same roots had caused them to fail and stop supplying that side of the tree. The missing massive branch was a direct result of the desire path. This was a powerful demonstration of two basic concepts, neither of which was new to me in theory, but Kevin was showing me how they joined together and pulled us into the tree's story. He revealed how our own choice of path can kill parts of trees.

After thanking Kevin for his time, I headed home to Sussex, joyous and excited. I had barely dropped my notebook on the kitchen table before I continued out of the back door and into the woods. I walked a short circuit through the trees, which I know very well, and couldn't believe what I was seeing. I was following a popular shortcut through the beechwood, a desire path, and every few seconds, I could spot another struggling branch. The dead branches were on both sides of the path, but always on the side nearest to it. How had I not noticed this before? Those dead branches must have hidden before my eyes a thousand times or more.

Now it is your turn. Take any opportunity you can over the next week to spot a well-worn shortcut through trees, a desire path that seduces many; most city parks have plenty. Look at the branches that grow over that path. It won't be long before you spot death caused by desire.

Perhaps we shouldn't follow these paths. Should we feel guilty when we do? I don't think so, for reasons I'll explain in a moment. But my first job is to help you see these things. We can't learn to read something if it's invisible in front of our eyes. It's hard to miss the dead branches we walk under when we suddenly realize we're part of that story.

We risk harming all nature when we walk through it, but I'm a firm believer that nothing is at greater risk than the invisible. Besides, by the end of this chapter, you will know how to walk over roots without harming a tree.

FOUR SHAPES

Roots are the prime movers of trees, but before they can get going they need to work out which way to head.

If a seed lands the right way up, the root emerges from the bottom and keeps going down before branching out. If it lands upside down, the root tip emerges from the top, grows up for a tiny distance then does a U-turn and starts heading downward. This is geotropism at play again; the plant's growth in response to gravity. Roots are shy of light and grow toward shade, a trend known as "negative phototropism" to botanists. Once the root tip has grown a bit, lateral roots start to emerge and these also know which way to head: away from the main root and downward.

Theophrastus was an ancient Greek philosopher and a man I would dearly like to meet on the other side one day. He noticed the large and little things in nature but was especially fond of clues. He wrote treatises on pure philosophy but also produced a work on weather signs and two on plants. A little over 2,300 years ago, Theophrastus noticed that each spring the roots start growing before the higher parts of the tree. This is logical: The tree won't get far without water and minerals, so it makes sense to get the flow of those going as early as possible. To this day, botanists struggle to monitor root behavior, so noticing these trends more than two thousand years ago was impressive and inspiring. Nice work, Theo.

Like all other parts of plants, the roots of a tree will grow according to a plan, dictated by their genes, that adapts to the world they meet. Each species follows its own unique scheme, but we can group them into four main types: plate, sinker, heart, and tap. The names sum up the tree's priorities for its roots: Do they spread wide and shallow, like a plate, or try to burrow deep, like the tap roots? (Here, "tap" refers to the verb: These roots try to *tap into* the soil deeper down.)

Whenever one of the tall beech trees in my local wood comes down during a gale, it creates a familiar-shaped hole in the ground. The tree lifts a broad, shallow area of soil, very like you'd get if you buried a wine glass up to its stem and pushed it over. Directly below the trunk there is a slightly deeper hole, but outside this area, the shape is wide and surprisingly shallow. Beech trees, along with firs and spruces, have a *plate* root system.

Plate

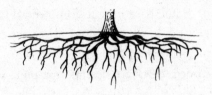

Sinker

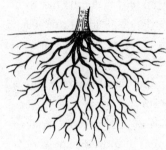

Heart

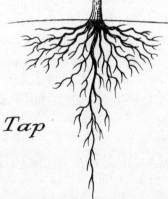

Tap

Some trees, including a few oaks, go wide then drop a few new roots vertically down from their lateral roots, which creates "sinker roots."

Birch, larch, and linden trees compromise: Their roots are quite broad *and* quite deep, growing into a shape known as "heart roots."

Some oaks have a deep central tap root when young, but it is less conspicuous in older trees. It is a more lasting feature in pine roots, which is why a storm that topples spruces can leave neighboring pines standing. The walnut is one of the few trees that retain a substantial tap root into maturity, which is one reason why it doesn't like to be moved around by whimsical gardeners. Walnut trees originate from central Asia and their tap roots make them more resistant to drought.

There's an art to understanding the shape of root systems in arid regions. If plants survive by accessing water from deep sources over long periods, they need a tap root. But if they rely on rain that comes in sporadic showers, as it often does in deserts, they need wide, shallow roots. Tap roots are rare in moist, temperate regions, like the UK and much of the US, but the few trees that have them, including the walnut, fared well during the very hot, dry summer of 2022.*

---

*I wrote about meeting the sprawling desert wildflower, *bindii*, in the Arabian desert in *The Secret World of Weather*. Its pretty yellow flowers were a sign that there had been a shower recently. As part of my investigations to understand how it survived in such harsh dry terrain,

The general rule is that roots can spread to approximately two and a half times the width of the canopy. Here's an even more general rule about depth: The roots of broadleaf trees are probably shallower than you would first guess. Most of the things the roots want, including nutrients and oxygen, are found near the surface and most of the work roots do takes place at a depth of only 2 feet (0.6 m).

There is a lot of debate about tap roots but, like politicians, they are more discussed than witnessed. People seem to like the idea that each tree sends a strong main central root deep into the soil. But when you look at uprooted trees, it is amazingly rare to see these famed tap roots. There are three good reasons for this. The first is that, as we have seen, most trees go for a wider, shallower root system. Second, we are most likely to see uprooted trees after a storm and plate roots don't cope as well with the strongest winds as the deeper shapes. Finally, tap roots are more important in youth than maturity. We can think of all trees having a tap root when they are a few weeks old and very few having one when they are mature.

Conifers have deeper roots on average than broadleaves. Firs and spruces have wide shallow "plate" roots, but most other conifers favor a bit more depth.

I learned two interesting things. First, it combines a tap root with a fine network of smaller roots. Second, the dried root is reputed to improve sexual performance, although the science doesn't support that theory. Weird science.

In the coastal area of Kalaloch in the Olympic National Park, Washington State, there is a Sitka spruce that has earned an affectionate nickname: the "Tree of Life." It is a tribute to the tree's determination to cling to life, despite nature making that almost impossible.

The Kalaloch spruce grew tall and strong, despite constant exposure to the coastal elements. It started well with good soil, plentiful light, and fresh water from a stream. Then, gradually, the fresh-water supply that had been a blessing became a problem: There was too much of it and it was too close for comfort. The stream that ran below the tree steadily carried the soil beneath it out to sea. The spruce found itself growing over a small gorge, held up only by the roots that had spread wide enough to bridge this gap.

The tree was fortunate to have a plate-root system: A narrow, tap-rooted tree could not survive a large hole opening directly beneath it. Like the fingertips of the hero in a movie clinging to the cliff edge, there was just enough strength in the roots on either side to hold the tree above the void. But the really interesting thing is not the shape or strength of the roots when the problem started, but their shape when things got tough.

Over time the stream carried more earth out to sea and the gorge deepened and widened until the poor spruce appeared suspended in mid-air. The trunk and all the major branches were now positioned over a hole that was almost as wide and deep as the tree's main canopy. Its response to this dire situation can teach us about the second important part of how roots grow.

The Sitka spruce's genes gave the roots a good general plan: Grow wide and shallow, like a plate. But the genes of any organism don't know what it will encounter: They offer instructions on how to get through a door, but not what to do on the other side. The growing parts of plants respond to stimuli and the roots copy the approach of the tree above ground: If they sense stress, they grow bigger and stronger.

The roots at the edge of the Tree of Life were placed under enormous tension, but fortunately not all at once. If the hole had appeared overnight, like a sinkhole, the tree wouldn't have had the strength to cope with it and would have disappeared into the abyss. The stream weakened the ground beneath the tree slowly enough that the stress on the roots grew steadily and allowed them time to build up their woody muscles. The roots at the edge of the plate are massively bigger and stronger than they would have been if the stream had not gouged the gorge. Some of the roots look more like major branches, and this is not a bad way to think of roots. They even have growth rings, like the trunk and branches.

Roots don't respond only to stress: They look for the good things in life, too. They will grow toward water and nutrients. And, like the top of the tree, they will fork and branch out if the leader root is decapitated.

Roots are quite easily blocked, but rarely dissuaded. If the tip meets an obstacle, it will make a small effort to go through it, but if this is unsuccessful, it will deviate by the minimum amount possible to go around the blockage and continue in the same direction. There is a widespread misunderstanding about the power of roots to go *through* things. Most roots aren't very good at tunneling through hard barriers, but they are very powerful when it comes to growing steadily thicker. This is where the myth probably stems from: We know roots grow outward and we know they are strong enough to lift road surfaces and paving slabs, but the two things aren't the same.

We now understand the two main influences that govern the form of the roots. There's the overall shape dictated by the genetic plan, plate, sinker, heart, or tap, and the environmental adaptations: Where have the roots grown bigger, stronger, or longer and why? It is hard to make out these patterns when the tree is standing, which is why we must take every opportunity we can to enjoy the forms we see on any tree that has fallen and lifted the roots out of the ground.

WINDS AND HILLS
The wind has a big effect on roots, but to understand it, we should start by reminding ourselves that winds

aren't random, however much it may seem that they are. Wherever you are in the world, the wind will blow from one direction more frequently than any other. In temperate regions, this is usually the direction that most of the strongest winds come from, too, and we call that the "windward" side of the tree. The opposite side is the "downwind" or "lee" side.

The roots have to deal with the effects of the wind and this leads to two opposing forces, our old friends tension and compression. Both forces lead to bigger, longer roots, but also different shapes. On the windward side the roots are like the guy ropes of tents: They are under tension. On the downwind side the roots are compressed: They act like props holding up an old leaning wall.

On average, roots on the windward side of the tree grow bigger, stronger and longer than other roots. This can be seen near the base of the trunk, where the roots flare out before plunging underground: a useful trend in natural navigation. We can use the thickest, longest roots as a compass: They point in the same direction that the prevailing winds come from.

The roots on the downwind side tend to be the second biggest, after those on the windward. This means you can normally spot a wider flare at the base on the west and east sides of trees in midlatitude areas of North America and Europe, and the roots stretch farther from the tree on these sides, too.

If you cut a root all the way through and look at it end-on, what shape will you see? Most people imagine that roots grow as long, thin cylinders and have a circular cross-section, like a hose. But each root is dealing with different stresses so the roots on each side are not the same shape. Roots on the windward side are under tension and grow in an hourglass or figure-eight shape. Those on the downwind side are compressed and grow into more of a T shape. It's obviously impossible to see this effect in buried roots, but do look for it in any uprooted trees.

Guy roots anchor the tree against the strong prevailing winds and can be used to find direction.

I have a habit of wrapping my fingers around the roots of any toppled trees I pass: It reminds me to think of the patterns above, but it's also a good way of sensing shapes my eyes have missed.

JUNCTIONS, BUTTRESSES, AND THE STEP

If the trunk traveled vertically down into the soil and the roots headed out horizontally from there, it would create a right-angle and a weakness. Whenever the wind blew, there would be huge stresses at this junction. That is why the base of the trunk and the roots meet at a curve just above the ground. It softens and shares the burden of these stresses. The effect varies with species, but the gentler the curve, the more strain on the tree.

From a little distance we won't see the roots, but the curve at the base of the trunk is still visible. It is normally more pronounced on one side of the tree— the windward side—so the base of the trunk appears asymmetric. It creates a shape that reminds me of an elephant's foot, where the elephant's toes point into the wind. Some species take this logic to extremes and grow "buttress" roots. The junction is replaced by mighty root struts that reach quite far up the tree. Buttress roots are more common in soft, moist ground and the trees that grow there—like poplars—and are widespread in the soggy-soiled tropics.

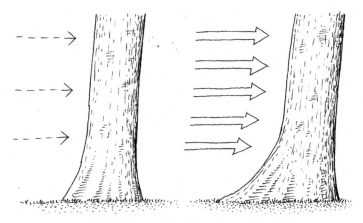

The "elephant's toes" point into the prevailing wind.

Roots adapt to gradients. Trees don't actually know which way is up or downhill and neither do their roots, which can only respond to the forces they sense. The roots on the downhill side will be under more compression; those on the uphill side will be under tension. On both sides the roots grow thicker and stronger according to the strain, but the lengths can vary wildly. Sometimes a tree can hold itself up from below by growing short, stout roots on the downhill side to cope with the extra compression and push against gravity. This cannot work on the uphill side as the roots need more length for leverage to hold the tree upright.

It's easier to grasp the logic if we give ourselves a similar challenge. Imagine your job is to make sure that a barrel full of water stands up perfectly vertically on a steep slope, using only pieces of wood. If you do this with blocks under the lower side, you might need a few short fat wooden ones under the base to prop it up. But if you had to do it from the uphill side, you'd need a long strut, attached to the ground at one end, to hold it up. The tree uses both strategies, which leads to the different root forms on each side of trees on hills. As always, conifers prefer to use compression and push from below; broadleaves favor tension and pull from above. The roots also display the different figure-eight and T shapes we saw above. On steep slopes, the roots are more exposed: They stand out, which gives us a better chance to observe these effects.

The different angles of roots on the up and downhill sides of trees lead to another pattern you can look for. When walking down a steep wooded slope, you will come across something I call the Step. On the uphill side of the tree, the roots stretch away from the base, near horizontally, and this creates a small flat platform. But on the downhill side, the roots plunge down, near vertically, creating a sudden small drop as we step from the uphill to the downhill side of mature trees. I have found the effect very noticeable when we're walking down a steep gradient and using the trees for support and balance. This habit leads us over the most uneven ground, a series of steps—it makes our arms do a lot of the work, which feels odd, but can be a welcome change on a longer walk.

The Step

TWO ROOTS

Tree roots have a life below the soil, at the surface and far above it. Their lives are reflected in the higher parts of the tree. We saw earlier how trampling can kill the branches on the same side of the tree as the damaged roots. What you will quickly notice, when looking for this, is that sometimes the tree suffers serious harm from minor foot-fall, but at others seems quite healthy despite a busy path very near to it. It can be confusing until we understand why this happens. Some species are more vulnerable than others, of course, but that doesn't explain the variation we see across trees of the same species.

We live near a town called Chichester that was once the Roman settlement of Noviomagus Reginorum. The

Romans built one of their famously straight and well-engineered roads from Noviomagus to Londinium and it left bold marks on the landscape that survive to this day. Some of the ancient road now lies under the busy A29, but on a short walk from our back door, we can step back in time and onto a broad, straight footpath that cuts a clear line through the local woods.

In places, this ancient path narrows, then twists and contorts as it weaves its way between and over the roots of beech, hawthorn, ash, elder, and other trees. The ancient Romans would not have approved of the sinuous and bumpy nature of the path these days, but I do: It is an enchanting track, shared by walkers, cyclists, dogs, sheep, deer, rabbits, and the faintest echoes of hobnailed leather boots falling in rhythm on smooth stones.

For many years it struck me as odd that such a well-worn path has not killed off the trees growing at its edges, or at least the branches on one side of those trees. The footfall is heavy in the narrowest spots and pedestrians pass so regularly over the trees' roots that it leaves them shiny in places. Yet there are big branches that pass, thick and healthy, above. How do the trees at the edge of this busy path thrive when others struggle, or die, under lighter footfall?

Some of the smaller trees, like elder, have evolved to thrive in busy areas and have roots that tolerate the compaction. But something simpler and more fundamental is going on for all trees: Not all parts of a tree root have the same role. The roots perform two main tasks: to support and to supply. The thicker parts of the roots nearest the trunk have a structural job, while the thinner, spreading tentacles farther from the trunk convey water and minerals.

The parts of the roots that are farthest from the trunk are most vulnerable, and minor trampling is likely to do more harm to the canopy than heavy stomping on the thicker, woodier roots nearer the trunk. Footfall doesn't normally break the sensitive roots, but it compacts the soil and causes cavitations—air bubbles—in them, both of which result in serious water supply problems.

If you stand so that you can touch the trunk of a large tree, you are standing over, or possibly directly on, the thickest part of the root; it is the toughest part and can withstand a lot of wandering feet. But if you step nearer to the edge of the canopy, you'll be standing over some of the most sensitive roots. This area is known as the "drip line" and the delicate roots that lie just beneath the surface here collect the rain that flows down from the edge of the canopy. If you want to nurture a tree with water or feed, this is the place to do it. If a path forms over this area, the branches above it will start to suffer, fail, and fall.

The ancient Roman road is now a path that passes over the thick, tough parts of the roots. Ironically it winds too close to the big trees to do them serious harm. They continue to thrive, even as the sound of ancient soldiers has been replaced by the clicking whir of mountain bikes.

A path doesn't have to be human to harm the trees. Many animals follow the same trails between food, water, and shelter. I once walked among the extraordinary elms of Preston Park in Brighton with John Tucker, an elm fanatic. He told me of a sheep trail he knew that could be read in the ground and in the trees: The path the sheep followed to their water trough had a line of dead branches above it.*

CRACKS IN THE SOIL

Messages are scribbled in the soil lying over the roots. If you spot especially dry soil near the base of any large tree, it is worth investigating. Dry soil cracks easily and we can use these lines to gauge how securely a tree is anchored. If there are more cracks on the side that the prevailing wind comes from, it's a sign that the tree is vulnerable to strong winds. If the cracks spread and form semicircles (with the trunk at their center), the tree may not survive the next big blow.

Trees in cities are more sheltered than those on wind-swept hills, but cities have their own strong wind characteristics. You may sometimes see cracks near trees

---

*I've spent a bit of time looking for something that I know must be there, but haven't found yet. If we know that the roots at the edge of the canopy draw up lots of water and those nearer the trunk don't, it means there must be "circles" of wetter and drier soil as we walk away from the tree. This should influence the color of the soil and plants we find there, as fluctuating moisture levels always do. But so far I haven't found any clear signs of this. Perhaps the extra water flowing down the edge of the canopy at the drip line perfectly balances the extra uptake by the feeder roots there, but that seems a bit too neat to me. The search continues.

in concrete, paving stones, or asphalt, and this is more common when trees are on long avenues that channel the wind or near very tall buildings that lead to big gusts.

SHALLOW ROOTS

I remember exploring a wild and wet part of Devon, in South West England. There were water-loving trees and lower plants all around me, including willows and rushes, and I could see lots of tree roots at the surface. These observations are connected.

All trees want their roots buried at least a little under the soil, even the plate ones. If we see roots at the surface spreading well away from the trunk, it signals that something in the soil is troubling the roots. Waterlogging is one possibility: If the water table is abnormally high, the roots are forced to rise over it. This isn't because they don't want the water, but because they need oxygen, and stagnant water is low in this vital gas. Conifer roots are more sensitive to waterlogging than those of broadleaves, but both feel the effects.

If water is the reason for shallow roots, the trees are doubly vulnerable to strong winds: Their roots are too near the surface, not well anchored, and the ground is soft. Shorter trees, like many willows and alders, cope with this because they stand well below the strongest winds, but it is interesting when tall trees survive with shallow roots. This is a sign that they are probably in a wind shadow, perhaps at the downwind side of a wood or in the lee of high ground. These trees are extremely vulnerable to the rare storms that come in from unusual directions. Storm Arwen was just such a storm, and in late November 2021, it swept in from the northeast, uprooting many trees that had stood in the Lake District for at least a century.

If you see shallow roots and water isn't to blame, it's a sign that the soil is thin, probably with rocks not far below. The roots have been squeezed up to the surface. If so, you'll see plenty of evidence of this in rocks among the soil and in its color.

Whenever you spot roots at the surface, spreading far from the trunk, look at the canopy. There's a good chance you'll notice that it isn't in full health. If the feeder roots are desperate and struggling, the tree probably isn't getting everything it needs for a glorious crown.

CLAUSTROPHOBIA

I was once standing at a pedestrian crossing at a highway junction in Texas, waiting for the signal to change. I had pressed the button and stood there hoping it would be just a few seconds before the light invited me to cross. The traffic flows changed once, then again and again, but still the glowing pink-orange hand showed that I was not allowed to cross. I grew impatient and walked along the highway in the direction I needed to go until I spotted a gap in the traffic then ran across to a narrow island between the four-lane highways.

Unfortunately, instead of solving half a problem, I had jaywalked into a bigger one. I was stuck on a concrete island between eight lanes of enormous fast-moving vehicles. I watched a little old lady motor past me in a pickup truck with monstrous wheels and an angry exhaust. I imagine Texas's oil history explains why the smallest car allowed on the road could pull a tractor. Anyway, the traffic flowed thick and fast for what seemed hours without pause. I was stuck with no legal or safe way of getting off my fume-soaked island. Resigned to starting a new life in the middle of a highway, I looked for things to distract me. There were some boring advertisements, and a much more interesting black vulture eating the remains of some unfortunate creature on the far side of the road. But then I hit the jackpot: a line of small trees, crape myrtles, I believe, growing in a narrow dry flowerbed set into my concrete island.

To pass the time, I walked along the single-file line of trees and noticed that they got taller, then smaller. The trees at either end were the shortest, the next ones in were a bit taller and the one in the middle was the tallest of the lot. They were all small trees, but there was a definite trend from short to taller as I walked from each end toward the middle. This was nothing to do with age: The trees were the same age and probably planted on the same day.

At first I thought this must have been a display of the "wedge effect" we met earlier—the trees at each end were shorter, thanks to bearing the brunt of the wind, perhaps some of it fanned by passing cars—but then I spotted the true cause. The flowerbed was not an even width: It was skinny at the ends and fattest near the middle. The roots of the trees at either end were tightly boxed in by concrete, meaning those trees struggled and grew less tall than the one in the middle with the widest section of flowerbed.

Eventually I escaped through a pause in the flow of trucks, but not before a line had started bouncing around in my carbon-monoxide-addled brain:

"A tree cannot bloom if the roots ain't got room."

Now that I'm seasoned to look for this effect, I find it pops up all over the place, including on rocky landscapes and regularly in cities. It's very common for urban landscapers to plant a tree where there isn't enough space for the roots. If you see a row of planted trees in a town and one, usually at an end, that is shorter than all the others, the chances are that it has claustrophobic, boxed-in roots.

CURVED FINGERS

I walked into a yew forest in Sussex and felt the dark, close atmosphere of the ancient trees all around me. It was dusk in winter and the owls added to the moodiness of the habitat. I walked on a few paces and shuddered as I saw the curved fingers reach out of the earth. I'd passed it many times before, but in the fading December light, the dark wooden hand made me shiver.

There is a strange tree pattern you may come across that confuses almost everyone who can't decipher the earlier history of the tree. When low branches touch the ground, the tree senses this and triggers a response: New roots sprout and sink down into the earth. They give the low branch an independent source of water and nutrients and it no longer needs the parent tree. Over time the link between the branch and its parent withers and disappears, leaving a new clone tree about 10 feet (3 m) from its parent. This process is called "layering" and it is more likely to occur if the parent tree is suffering in some way.

The new tree always has a peculiar shape because it started life as a horizontal branch. However mature it grows, there will always be a signature bend or kink near the ground. Sometimes several of the parent tree's branches touch the ground and the process is triggered in all of them, leading to a ring, or more commonly an arc, of new trees. When this happens the young trees always start with a radial pattern, growing away from the parent tree, in the same direction that the branches were heading when they touched the ground.

It is fairly easy to spot the recent history of layering when the parent tree survives. We can still see some low branches dipping toward the ground and imagine the rest. But in long-living species, including yews, the new trees will often outlive their parent by many years.

This is what creates the strange pattern: a partial ring of curved trees, like fingers reaching out of the soil.

If you see a peculiar ring of trees that started life near horizontal, then curved upward toward the vertical, look to the center of their circle. You may spot a decaying stump or another remnant of the parent tree.

FLOTSAM PATTERNS

Interesting patterns form around the base of trees. As they stretch away from the base, roots form a net ready to catch anything passing on the wind. It carries dead leaves, dust, small twigs, feathers and more across the ground. The trees get in the way of "wind flotsam," some of which falls out of the breeze and starts to pile up at the base of the tree. There are always patterns we can look for in the little piles we see around the roots.

Dead leaves are the flotsam we will see most regularly in or near woodland. If you study the hollows between the guy roots, you'll start to notice that the leaves have preferences: They tend to gather in small deep piles on one side of the trees, while the opposite side has hardly any. On slopes you'll see more leaves gathering on the uphill side (which compounds the Step effect). On flatter ground, patterns are formed by the wind, with two aerodynamic reasons for them.

Anything carried by the wind is likely to fall out of the breeze if the air is slowed down. Whenever a wind

Flotsam Patterns

meets an obstacle, there are places it can't reach, and, as we saw earlier, these sheltered spots are known as "wind shadows." They act like magnets for dead leaves, which fall into the still air and rest there because no wind can reach in to whisk them away. They pile up in that one tiny sheltered spot.

The second reason this happens is that the shape of the roots is different on each side. On the windward side, the roots point out more, while on the downwind side they jut out only a little, and this forms wider, deeper hollows for the leaves to rest in.

The other day I crossed a wooded hill that segued from beech to oak and then western red cedar. The leaf litter changed as the species did, but the patterns under each tree were consistent, conspicuous, and easy to navigate by. Lots of small piles of leaves were tucked in on the northeast side of the tree roots, but few on the southwest. It's up to us to decide whether or not to use these patterns to find our way or just to add a little pleasure to the path we are following.

How to See a Tree

AN INTERLUDE

A FEW DAYS AGO, I spent some time in a small circular wood on top of a hill not far from home. While I was there I spotted a pattern in the trees that was so simple, elegant, and practical that I was shocked I had never noticed it before.

We will meet this pattern later in this chapter, but this is a good opportunity to consider, briefly, the science and art of perception. We can't read something we haven't spotted. And, as I have proved to myself many times, it's not as easy as it sounds. The keys to noticing things are hidden in the following simple story.

I decided to wander around East Sheen Cemetery in South West London. I had time to kill, so why not spend it among the dead?

A small tidy square of perhaps a hundred graves opened through a gap in a hedge to reveal thousands across the grass. They lined up neatly, row after row, pale rectilinear stones stretching to distant boundaries. Each one was an individual, each one lost in the mass.

A pale blue, very old Ford Escort passed between two rows and stopped. It startled me. I'd had no idea a car could reach that far into the cemetery. An elderly lady stepped out, white hair, dark coat, clean, tidy. There was a familiar careful rhythm to her movements, one leg out, then the other, hand on door, up we go. Worried she might sense my eyes on her, which could spoil a private moment, I turned away and walked in the opposite direction. There is a feeling of trespass when you know none of the dead in a graveyard.

In enclaves, there were many trees, not as many as the graves, but many more than I'd expected. And variety: yews, of course, but cedars, pines, birches, oaks, and maples, too. It struck me how many were in pairs or groups of three or more. A pair of yews, a pair of maples, then four planes; neat and measured variety, a symptom of human curation.

There weren't many of us about: The few visitors were comfortably outnumbered by the gardeners and their noisy machines; high-pitched gas engines were trying to wake the dead. My friendly "Hello" to a couple was ignored and I soon stopped offering even a smile to those

I passed. I adopted the code of the place from the others: We didn't look at each other, we didn't greet each other. Each to their own.

I walked toward a silver birch whose branches swayed in a way that intrigued me. And then I couldn't help but glance at a group of three women, one middle-aged, two very young, possibly still teens. They were arranging flowers in vases on a trestle table. And laughing.

"Not sure what there is left to do now," the eldest said.

"No. Pretty much there, then it's time to . . . party," one of the young replied. There was a peal of laughter, stifled, perhaps, because I was within earshot.

"And shop," the third added. More laughter.

They were clearly preparing for some informal memorial or similar event, but couldn't surely be celebrating the death of a relative. It didn't make sense. The urge to close the distance between us was strong—I desperately wanted to know more. But it wasn't my business and burning curiosity couldn't make it so.

The things we notice stand out in some way. Motion attracts our attention, which is why many prey animals, like deer and rabbits, will freeze if they spot us. I didn't see every tree in the graveyard, but I had noticed the birch's branches sway.

Anomalous shapes, patterns, and colors force their way into our thoughts. The pale gravestones formed straight lines that stood out against the grass. I had not expected to see that old car, in that old-fashioned pastel blue in the middle of a graveyard. It moved and it jarred, so I couldn't help but notice it. Contrasts stand out—the lady's white hair against her dark coat.

I made a lot of minor observations in that unremarkable but true short story. I want to focus on some that are easy to miss.

Perception has two parts: the physical and the psychological. We can improve our physical observation powers using lenses, from telescopes to contacts. We can also improve the psychological side, and one of the best ways to do that is by enhancing our motivation. We notice more when we care about what we're looking at.

There are personal aspects to this: We notice the faintest smile or sadness on the face of someone we love. Evolution has programmed that into us. (We are also fine-tuned to understand the motivations of others, which is why eavesdropping can be tempting.) There are pragmatic reasons behind strong motivation: If we are using nature to navigate or forage for food, certain useful patterns will shine out, especially if our lives depend on it. But the fascinating thing is that we can actually enhance our motivation in situations where there is no burning personal or pragmatic agenda.

Something truly magical happens when we learn just enough to expect to find meaning. The American behavioral economist George Loewenstein has put forward many groundbreaking ideas, including the Hot-Cold Empathy Gap. It states that we are bad at understanding a frame of mind when we are in a different one. It is hard to imagine being too hot when you're too cold, or too hungry after an all-you-can-eat buffet.

In the early 1990s, Loewenstein proposed one explanation for curiosity. Scientists, even social scientists, like researching things that are easy to define and measure. Money ticks both boxes, and every economist studies it, but curiosity fails on both fronts. There are many fewer studies on the causes and consequences of curiosity than there are on the effects of saving money. Even though any sane person would argue that curiosity shapes the world in much more intriguing ways than savings accounts. The Information Gap is Loewenstein's brilliant attempt to right the imbalance. Curiosity is:

"a cognitive induced deprivation that arises
from the perception of a gap in knowledge and
understanding."

Or curiosity is an itch. We feel it when we have some information but sense more is missing. If that doesn't sound groundbreaking, it may be because we tend to focus on the *missing* part when the originality lies in the bit we *do know*. You can't have a gap without the stuff on either side of it. The implications of Loewenstein's work are powerful—they highlight that knowing something makes us more curious than knowing nothing. Some information lights the fuse of curiosity in a way that total ignorance doesn't.

The great news is that we can create gaps—we can engineer our own curiosity. The trick to making ourselves curious about any puzzle is to fill in some of the blanks. When you've nearly finished a crossword and only two words are missing, it is a more compulsively curious situation than the blank puzzle.

Every single time we see a tree, we can fill in a lot of the blanks quickly and easily: its shape, its colors, its leaves. And we can fill in another part equally quickly: by spotting that these differ in some way from the norm. There simply must be some reason for those differences. This is the gap in our knowledge. It is always there, and once we know that, we will always find it. It lights the fuse of curiosity. We learn to look for these differences and we can't help wondering what they mean. And then we see the tree properly for the first time.

The extraordinary observation I mentioned at the start of the chapter was in the shape of tree roots. In the previous chapter we looked at lots of root patterns, including the way the roots grow bigger, stronger, and longer on the side of the prevailing winds.

That afternoon in the circular wood, two things happened that allowed me to see something that had hidden in front of me for decades. The sun was behind clouds for most of the afternoon, but then it dropped below them and shone its orange light onto the wood I stood in. The canopies of the trees shaded much of the ground, but the light caught the lowest parts of some of the trees in front of me. It created an interesting effect highlighting the area where the trunk flared out to the roots. The contrasting colors and the motion created by the bottom of the clouds passing over part of the sun drew my attention to the roots. I noticed they pointed strongly in one direction.

My brain wanted to leap to the conclusion that they were pointing southwest, as I would have expected them to, but it took me just a moment to appreciate that that could not be true—the sun gave me a strong sense of direction and wouldn't let my mind leap to such a false conclusion. The roots pointed north.

How odd, I thought. What's going on? There was now a gap in my understanding and a burning curiosity to decipher the meaning of a casual observation. The feeling was strong enough to keep me in the area for half an hour, all the time looking intently, until I solved the mystery— until I had filled the Information Gap. Motion, colors, contrasts, and now curiosity helped me to see things I might have missed.

After spending time looking very carefully at the tree roots at the edge of that small, circular wood, I noticed a new pattern: They all pointed to the edge of the wood.

It suddenly made perfect sense. At the perimeter, winds are stronger on all sides of a wood than they are in the center. Tree roots grow bigger and longer on the side the strongest winds come from: Of course the roots near the edge of the wood on a hill will point out to the edge. The roots show the way out!

This book is about understanding the things we see in trees, but that is part of a virtuous circle. The more we know to look for, the more we look and the more we see. As this happens, we start to see some things we didn't look for, and they pose their own questions: We find them curious.

The great joy is that, however bizarre our route to noticing something new, once we have spotted it and deciphered its meaning, it shines out the next time and for evermore. It can no longer hide.

9

Shape-Shifting Leaves

Size Is Something • Shape-Shifting • Do Not Compete Against Yourself •
The Reverse Tick • Growing for Flow • Hotter and Higher Lobes •
You've Changed • Shimmering in the Dry Light • Light and Dark Greens •
Blue Delight • Yellowing • Obvious and Invisible • White Lines • I
Feel Your Pain • Stalking Leaves • On Maneuvers •
The Conifer Net

IN ANCIENT GREECE, when people had a difficult decision to make, they sought guidance from priestesses known as oracles. Two of the most famous were the oracles at Delphi and Dodona. Pythia, the oracle at Delphi, was famous for her cryptic and near-nonsensical ramblings. One theory is that she was intoxicated after chewing laurel leaves or inhaling their smoke.

The oak was sacred in ancient Greece, as the tree of Zeus. When travelers reached Dodona, they would look for the priestess who slept under a very special oak. The oracle would listen to the traveler's dilemma, then turn to the oak for signs and find them in the rustling of its leaves, which was believed to be the voice of Zeus.

There are meaningful signs in the leaves of trees and in this chapter we will learn how to interpret them without the need to consult high priestesses.

All tree leaves are trying to perform the same simple tasks: They need to harvest sunlight and exchange gases as efficiently as possible. Given that they all have the same two jobs, it is remarkable that there are so many different forms. The sun doesn't change significantly and neither do the gases, so why is it that we can see fat leaves, skinny needles, ovals, triangles, lobes, teeth, spines, wrinkles, dullness, shininess, long stalks, short ones, simple patterns, and complexity on the same short walk? We can notice a million things about the leaves on a tree and all of them hold some meaning; the key is knowing which features carry the most interesting messages.

Nature is not a whimsical artist splashing variety onto our landscape in the hope of winning prizes for creativity. There must be a reason for every one of the differences we see, and once we have it, we see the sign. Trees experience wildly different levels of water, wind, light, and temperature over small distances and the leaves reflect these changes back to us.

The next time you walk down a street, try to notice whether other people are walking with their elbows tucked in or held away from their bodies. You rarely see people with their arms out or raised when the weather is harsh. In cold winds, animals shrink themselves, tucking their limbs in and making their bodies more compact and less vulnerable to heat loss. Conifers have small needles or scales because these compact sizes cope better with harsh environments. Broadleaf trees grow smaller leaves in cold or exposed areas. As a general rule, the more exposed to wind or cold a leaf is, the smaller it will be.

A tree growing in a windy spot will have smaller, thicker leaves than it would if it grew in a sheltered place, but we can look for this on a smaller scale, too: The most exposed parts of an individual tree will have smaller leaves than the most sheltered parts. If you pass two trees of the same species on a walk, one on top of a hill and one in the valley, the leaves at the top of the higher tree are likely to be the smallest and thickest; those at the bottom of the tree in the valley are likely to be the broadest and thinnest.

Leaves also respond to light levels. Trees have two main types of leaves: sun and shade. Sun leaves are smaller, thicker and a lighter color. We find more sun leaves nearer the edges and top of the canopy and on the sunnier southern side. The wider, thinner, darker shade leaves are more common low down, on the inner parts of the canopy and on the northern side of the tree.

Leaves respond to their world. If a tree is shaded by a new tree or building, the leaves will change from sun to shade: they will grow wider, thinner, and darker. This ability for leaves to morph according to changes in their environment is called "plasticity." The "decision" to change from one form to another is made when the tree forms new buds near the end of the previous growing season, ready for the start of the next.

Leaves are smaller on average in dry areas: Big leaves are prone to overheating and smaller ones are better at conserving water. You're more likely to see large floppy leaves in shady wet regions and you'll find the biggest in jungles.

Bringing the pieces together, we will find abnormally small leaves or needles in areas that are sunny, dry, cold, and windy. Study a leaf near the tree line—which marks the top of woodland on any mountain—and compare it to one near the river in the valley and you'll see a staggering difference in size.

SHAPE-SHIFTING

Some places are much windier, darker, or wetter than others, so it's easy to see why there is such variability in size. But why such wildly differing shapes? Botanists reach for a rich smorgasbord of terms to describe leaf shapes, some that almost paint a picture—ovate, deltoid, rhomboid—and many more obscure ones that bring a smile and not much imagery, such as cordate, bigeminate, imparipinnate, or palmatipartite.*

An interesting relationship exists between leaf size and branching pattern that we looked at earlier from the branch end. It is worth revisiting it from the leaf end: The smaller

---

*Ovate—egg-shaped, oval; deltoid—triangular, shaped like Greek letter delta; rhomboid—diamond-shaped; cordate—heart-shaped; bigeminate—having two leaflets, each of which has two more; imparipinnate—with an odd number of leaflets and a terminal leaflet; palmatipartite—having palm-shaped lobes with deep incisions.

the leaves on a large tree, the more branches you will be able to count. The reason is straightforward: You need one stout branch to hold a giant leaf, but you need lots to fill in the gaps with tiny leaves. This is obvious in hindsight, but it's exactly the sort of thing that hides in front of our eyes unless we choose to notice it. It is a pattern you can enjoy exploring closer to the ground, too: Notice how giant-leaved plants, like rhubarb, have a single thick stem under each leaf, but herbs with hundreds of tiny leaves divide into lots of mini-branches.

When looking at leaf shapes, the first big call we need to make is whether we are looking at "simple" or "compound" leaves. Is the leaf you're looking at an individual (simple) or is it part of a gang of leaflets (compound)? Simple leaves are attached by a stalk, a "petiole," to a bark-covered twig. Compound leaflets sprout off a green rib, a "rachis." If pairs of leaflets run off the central green rib, this is called "pinnate compound."

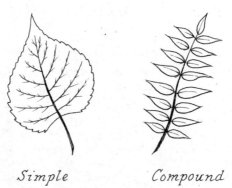

Simple   Compound

In cool temperate regions, pinnate-compound leaves allow a tree to throw out lots of leaflets and harvest lots of light quickly, without the slower, expensive growth of a woody twig. This arrangement suits fast growth but works better in places that are bright or windy. These strengths go together: If there is an area of open land, a pioneer tree will want to grow rapidly, but it will have to contend with lots of direct sunlight and little shelter from the wind. The light also filters through the gaps in the compound leaves, meaning that the lower layers in multilayer trees can also harvest some of the light.*

In summary, pinnate-compound-leaved trees, including ash and elder, are a sign that a tree has made the most of a gap: It has seized an opportunity. And, as is always the case with pioneers, this means the tree landscape is still young and will change dramatically over the coming decades.

DO NOT COMPETE AGAINST YOURSELF

I remember enjoying a family picnic on a blanket in a London park a couple of years ago. We were sitting near the paved path that ran around the edge of the park, and panting runners passed us every few seconds. I don't know about you, but to me, exercising is good if I'm the one

---

*In the hotter, drier parts of the world, compound leaves mean something different. They allow trees to shed the whole rachis in times of drought, effectively ditching bunches of leaves to conserve water.

doing it, but it looks a little ridiculous when I'm in a different headspace or my mouth is full of egg sandwich. One of the runners who passed us was wearing a T-shirt with ME vs. ME on it. I had to stifle laughter until they had passed. When interviewed, some professional sportspeople like to claim they are not competing against other individuals, only themselves. This is patently absurd and nonsensical, but it is a popular way of deflecting unhelpful questions.

All organisms compete for resources and often with others of the same species. It's tough enough out there already and the very last thing any plant needs is to add an internal fight. If a tree falls down in a wood, the two neighboring trees will compete with each other and any pioneer upstarts for the new light. They definitely can't afford to compete with themselves as well.

Imagine if every branch and leaf on a tree grew toward each other in a bid to grow above one another and shade each other out. This is not a strategy that would last long. Leaves need a lot of energy to grow: There are no spare resources to grow leaves over one another on the same branch. They have to cooperate and they do this by following a plan.

The simplest plan is for a branch to hold out the leaves in a flat plane, like a plate. If there are no branches above the wide flat plate of leaves, there is no danger of shading. Beeches and maples favor this arrangement, but this is an idealistic scenario that only works in the long term for monolayer trees. Many trees can't count on that plan and have to grow some leaves higher than existing ones on a lower branch. A few neat strategies allow for cooperation rather than competition, and the most popular involve changing the angles or lengths of the stalks.

If the plant makes sure that the leaf above grows at a different angle from the one below, the danger of shading is reduced. Seen from above, this leads to an effect a little like a spiral staircase in which each step is a new leaf. The strategy is easier to spot in lower plants than trees. The next time you pass a small leafy shrub or herb, try looking down on it from above: You'll soon spot how the plant achieves two clever things. First, you don't see much of the ground because the plant has covered most of the open area with its leaves. Second, if you imagine drawing a vertical line from your eyes to the ground, it doesn't pass through more than one leaf, because the plant has arranged them to avoid inefficient doubling up.

There is another way for a tree to put out a leaf without shading the one below it. If the tree shortens the stalk so that the higher leaf is closer to the stem, it won't cast a shadow directly onto the lower leaf. The higher leaf is also smaller, which is logical—it would be daft to have the larger leaves on top. Plants are not daft.

These strategies are much easier to see near the ground than high in the canopy, so whenever you see leaves on branches near the ground, make sure to pause and look for the plan that allows the leaves to cooperate. Is it a wide plate, clever angles, shorter stalks, smaller leaves on top, or an ingenious combination?

THE REVERSE TICK

As we have seen, the branches on trees grow toward the light, leading to the "tick effect" (or "check mark effect")— branches on the southern side are closer to horizontal; those on the northern side are closer to vertical.

Leaves have their own tick effect, but it's reversed. The leaves on the southern side of trees are closer to vertical: They point toward the ground. Those on the northern side are closer to horizontal. The reason is simple: Branches grow toward the light, but leaves orient themselves perpendicular to the light because they need to face it to harvest it. More light arrives from low down on the southern side, but on the northern side most of the light comes from above.

It can be a little confusing, but it may help to imagine a strange scenario. Let's pretend we're sitting in a tree, hugging the trunk, and we have a small solar panel in our spare hand. It's dark near the trunk, but we want to harvest as much light as possible. What is the best strategy?

It depends on which side of the tree we are. If we're on the southern side, we sense the sun out on that side and head to it (horizontal branch growth). Then, when we reach the canopy edge, we hold our solar panel (leaf) so that it faces out to the sun (closer to vertical).

If we started on the northern side we won't sense any light from the southern side—too much tree in the way—but we will sense some light high above us. So, we need to climb upward (vertical branch growth); when we sense the light is bright enough above us, we hold out the panel flat to catch that high light (closer to horizontal).

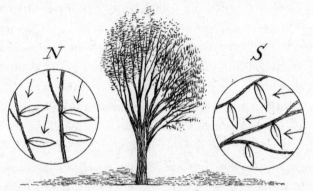

The "Tick Effect" in the Branches and "Reverse Tick" in the Leaves

This effect runs through every broadleaf tree and the lower plants, too. Occasionally we will spot a leaf oriented in a way that doesn't seem to make sense and this is a great opportunity to pause and remember something important. The leaves don't care about north or south: They care about light. Just as we will find branches growing out toward rivers and into avenues regardless of aspect, we find that leaves face the lighter zones, too. And whichever side of a wood we walk toward, the broad leaves will face you: We always arrive from the light.

There are so many species of willow and so many hybrids within them that it isn't much fun trying to identify or name them perfectly. However, the shape of their leaves is interesting and easy to read. There's a species of willow I see every day when walking near home that most don't instantly recognize as a willow because its leaves are oval and broad, for a willow anyway. It's called the goat willow. I see other willows, including the crack willow, less frequently, which have the classic willow leaf shape—long and very thin. It's called "lanceolate" or "lancelike."

Without any need to nail the names, the willow leaves I see follow a simple rule: The closer to moving water I find them, the skinnier their leaves tend to be. The oval-leaved goat willows are found in damp soil, but hardly ever next to moving water. The crack willow and other skinny-leaved examples are telling me that I'm nearing the river's edge. Thin leaves cope with flowing water much better than broader oval shapes.

Alders grow near flowing water, but have broader leaves, which seem to break the logic. But this is another clue. Alders and willows both grow at the water's edge, but they follow different strategies. Willows expect to lose the fight with a fast stream and can actually turn it to their advantage. When flowing water breaks the thin, weak twig of a willow, it ferries it downstream until it catches in the mud of the bank, where the tree can start life again from this natural cutting. A single willow can propagate downstream, which is one of their strategies for lining the banks of rivers for long stretches. Willows lose the battle but win the war.

Alders take a different approach: Their trunks and roots are stronger and built to resist the flow. They not only hold their own, they actually defend a riverbank against erosion. But there is a limit to this approach, which is why alders thrive by gentler streams and especially in wide areas of still or slow-moving shallow water, where they can form wetland woods known as "carrs." Alders also hold their leaves well above the water, unlike willow leaves, which may touch the water or hang a hand's width above it.

There are willows and an alder carr in a wetland area in South East England I love to explore: Amberley Brooks. One summer afternoon, I was wiling away a pleasant half-hour, sitting on some logs on a dry raised patch near the carr, staring out into some alders, waiting for others to join me for a walk. It was a period of enforced nonaction and such moments are good for the mind. The others were a little late, but it didn't bother me: The weather was fine and my mind was busy, playing with the difference in the alder and willow leaves. By the time the others joined me, I had a pair of lines in my head.

The leaves on the alder are taller.
Those on the willow grow low.

Many trees have lobed leaves, and five fingers, or lobes, to each leaf is the most common pattern, especially in the maple family. Lobes are pretty, but nature can't afford pretty for pretty's sake, so why do lobes exist? Lobes break up the edges of the leaf, which changes the airflow over and around it, making it easier for a leaf to shed excess heat. Lobes act like a fan on a hot day. Trees with lobed leaves have deeper, more emphatic lobes in their sun leaves, so we can expect to see more emphatic lobes higher in the tree and on the southern side. By deeper lobes, I mean that the indentations are more noticeable. We can think of it like this: If we trace a line around the edge of a leaf, the lobes mean there will be times when we take a short detour toward the center. The closer we get to it, the more deeply lobed the leaf is.

One pleasant December morning, I spent a few hours on Kensington High Street in London. The day was slightly ruined by the need to do some Christmas shopping. I enjoy spotting a gift and giving it, but it's the shops I struggle with. I find them exhausting. After locking horns with three shops and buying some gifts, I felt a bead of sweat on my forehead and rewarded myself with a pause in the cool outdoor air.

I stood on the broad pavement and breathed the tired breath of the reluctant consumer. Glancing into the window of an upmarket bakery, I could see cupcakes with more icing on top than sponge below and a price tag that I felt might melt the icing and set fire to the sponge. Depressed by this sugary robbery, I looked away and found my eyes resting on a pair of Oriental planes.

The Oriental plane has deeply lobed leaves, and as I looked from the lowest leaves to the highest, I could see the lobes growing deeper and deeper. Near the top of the tree, the lobes were exceptionally long and the leaves looked like the feet of some five-toed monstrous bird. It was very satisfying to notice this shape-shifting in the leaves and cheered me more than any expensive cupcake ever could.

YOU'VE CHANGED

There is an old saying: Generals always fight the last war. Sometimes head teachers do, too. When I was at school in my teens, the rule for hair length was simple: It mustn't touch the shirt collar. The school thought they had come up with a simple way of stopping us repeating the sins of our parents' generation, riddled as it was with hippies.

But we were teenagers and we knew that rules were for bending. We grew our fringes until they were long enough to touch our chins. Then we pushed the fringe all the way back until it almost reached the collar at the back of our neck. As soon as school finished for the day, we'd shake our heads and express ourselves through our unkempt mops of hair, like the exuberant, rebellious idiots we were. My sons' generation has decided that the "mullet" style needs resurrecting, and they are all busy cutting it short at the front and sides and leaving it long and ragged at the back. It is a hideous development, but . . . to each generation, their folly.

There are few certainties in life, but our appearance changing as we mature might be one. And the same is true of many leaves. If you look at a collection of trees of the same species, you might notice that the leaves you see on the older trees look different from those on their younger neighbors.

Some plant species have different juvenile and adult leaves: The form of the leaf changes between the young and the mature stage of a plant's life. Scientists are still researching why this happens. The most logical theory I have come across is that young trees are builders and need lots of carbon; older trees are survivors and need to cope with a long life in the elements. The leaf shapes morph to suit these priorities. Researchers discovered that environmental stresses, like hot or cold spells, can help trigger the change in plants from juvenile to mature—what doesn't kill the tree makes it more mature.

A few trees change so markedly that we may not recognize the tree at all from one leaf type if we didn't know it: Eucalyptuses, for example, have leaves that morph from rounded to long and thin as the tree matures. But there is no need to hunt out individual species that display extreme changes: It's more satisfying to look for the distinct but subtler changes all around us. The leaves on young conifers nearly always look and feel quite different from their older neighbors. It varies with each species, but young conifer foliage is typically shorter, thinner, softer to the touch and more brushlike than that of older trees.

The fun catch in this game is that not all parts of a tree can be considered the same age. We are conditioned to think of people as having a single age: They might be a one-month-old baby, forty or ninety years old, but we'd never think of anyone as being all three at the same time. But this single-number age is a cultural convenience because, in a sense, they are all three ages: Their fingernail cells might be a month old, their heart cells forty years old and their eye cells ninety years old.

When we look at the top of a sapling, we are seeing the very youngest part of the tree, and as we scan down, we move back in time until we see the oldest parts nearer the bottom. Conifers put out juvenile leaves at the top of a sapling and more mature leaves lower down or on side branches. We also find the youngest parts of the tree furthest from the trunk, at the end of each branch.

Sometimes you will spot juvenile leaves nearer the trunk and mature leaves nearer the edges of the canopy. This will reflect the age of those leaves, but also that life is more stressful at the edges, which triggers the change. Anything traumatic that forces the tree to "start again," like coppicing, will lead to juvenile leaves, not mature ones, however old the parent stump might be.*

SHIMMERING IN THE DRY LIGHT

It takes time to appreciate that the many shapes, patterns, and colors we see in leaves are all a reflection of the micro-world in which they grow. It is a truth that leads to a simple prediction: We should expect to see the same trends appearing in parts of the world that share similar environmental conditions.

During walks in southern Spain, Greece, and Australia, I have seen many different trees, most of them under a fierce sun. They may appear at first to have very little in common, but as soon as we start to notice a similarity, it is hard to miss.

Olive and eucalyptus trees are native to very different parts of the world, but both thrive in hot regions. Olives

---

*You'll see this effect in lower plants as well as trees. I see it daily in the common climber, ivy (English Ivy, *Hedera helix*), a plant that has taught me so much about botany. The juvenile ivy leaf has several lobes and points and looks very different from the mature leaf, which has only one point. When leading a walk, I sometimes use this difference to play a trick. I wait until nobody is watching me, then pluck a young and a mature leaf from ivy growing up a tree trunk, then walk on a bit before showing both leaves to someone, one in each hand: "Can you identify each of these leaves?" Some can, but the most common response is for the victim to look perplexed, point to the darker, multi-lobed immature leaf and say, "That's ivy, but not sure about the other one."

flourish in the hot, dry climates of southern Europe, and eucalyptuses are dominant, where trees can survive, in the hot, dry parts of Australia. These two tree families have evolved to cope with intense heat and sun in different hemispheres, and although they have many different features, they share a silver tint in their leaves. It is a color that reflects some of the sun's light and makes their hot homes more habitable.

LIGHT AND DARK GREENS

You will have noticed that many trees have leaves of different shades and sometimes even different colors on their upper and lower sides. The effect can be seen on most trees and puts on a show in some broadleaves on a windy day. In some species, like the white poplar, it is striking enough to give the tree its name. (The white poplar also has a deeply lobed leaf and there is no surprise that it is hot and dry in its native regions, like Morocco.)

The upper and lower sides of leaves appear different because they have different roles. Most direct light hits the upper surface of the leaf, so this is where most of the green chlorophyll needed for photosynthesis is concentrated. The lower side is more important for gas exchange.

Green means chlorophyll, but there is chlorophyll and chlorophyll. Or, to be more precise, there are different types of chlorophyll and they vary in shade from light to darker green. The type of chlorophyll each leaf contains changes depending on its role. Leaves that are adapted to lower light levels and older leaves have more of the darker green chlorophyll. This is one reason why shade leaves are darker green than sun leaves and why leaves grow darker as summer progresses.

BLUE DELIGHT

My work involves more road time than I would choose, but I have learned a very simple technique for making this more positive than it might be. As usual, I had to learn it the hard, stupid way, but the lesson did stick. And that in turn led to my discovering a delightful tree clue.

I set up my natural navigation school in 2008 and knew it would be a struggle. Even friends and family who wished me well did so with eyes that revealed they thought it a professional gamble that couldn't possibly work. So, like almost every other individual who goes against common wisdom, I harbored a quiet determination to prove it could.

One of my simple philosophies was to say yes if I was asked to do something. Early in my venture, an invitation arrived in my email inbox to speak to a very small group, hours away in the north of England. I checked my diary, which had plenty of blanks, and said, "Yes."

A couple of months later I discovered a problem. It makes me laugh now, but I hadn't got the hang of managing my diary. I was free at the agreed time of 8 PM on the date in question, but I had since accepted a different invitation to work the following morning at 9 AM in deepest Cornwall, eight hours' drive away from the evening appointment.

After the first talk, I drove down that night, stopping every few hours for a nap, and got both jobs done. I earned less than the cost of the fuel, but as is always the case in any new endeavor, I learned a couple of valuable lessons instead. The first was that the number part of a date in a diary doesn't tell the full story of that day. The second was that it was idiotic to have to race away from new destinations because of appalling planning. If I had just a couple of spare hours at the first stop, I could have spent them exploring a wonderful wild landscape at no extra cost.

Ever since, I have made it a principle to try to find at least a little time at every destination and at places en route. It has proved to be one of the most worthwhile habits of the past decade. Now, when work crops up at the end of a long road, I quite often squeeze in a chance to micro-explore somewhere on the way there, somewhere near the destination and a third place on the way home. This small habit has led to more serendipitous discoveries than most of my planned journeys put together. So it was that I discovered the blue-tree compass.

One November day I was on my way back from a job in Galloway, Scotland, having chosen a less-than-direct route through some mountains that I'd never visited in the Glenkens area. I parked and prepared to venture off into the hills for a natural navigation challenge. These shorter journey-break challenges tend to follow a simple framework: Walk for as long as feels right, then let nature guide me back to the car by a different route.

I was getting my bearings before setting off, when the blueness of a spruce tree grabbed my attention. I'm sure we've all enjoyed noticing the faint blue sheen that covers some conifers from time to time—it so often pairs with the pleasant whiff of conifers in sun. But there was something very striking about this specimen: It was remarkably blue. When we look at conifers and say we see blue, we often mean blue-green, but this tree stood out because it was more blue than green, to my eyes at least.

I paused to admire it and walked around it. It was at the southern edge of a coniferous plantation, and as soon as I took a few steps, I noticed the tree looked less blue. At first I thought it was the shift in light: The sun was peeping out from between clouds, and I felt that its angles would have had a big impact on the hues I saw. That was true, but there was a true blue color that transcended that effect and painted only the southern edge of the southernmost of the trees.

I didn't know it at the time, but the blue color we see is caused by wax. I was looking at a thicker layer of protective wax on the needles, a layer that protects them from a dangerous type of UV light from the sun. The wax is thicker on the needles on the sunnier southern side of these trees, meaning their south side is bluer. Ever since this small discovery, I have enjoyed looking for the blue-tree compass. Any journey that yields a joyful discovery like that will always feel worthwhile.*

YELLOWING

As autumn approaches, trees draw chlorophyll from their leaves back into themselves: They dare not waste such a valuable resource. The popular colors we see in the leaves at this time of year, the yellows, oranges, and browns, are the colors of the leaves minus chlorophyll.

Sometimes you will see leaves that have turned yellow long before autumn. They are crying out for nourishment. The yellowing, known formally as "chlorosis," is a sign that the tree is lacking in one or more of the key nutrients, like nitrogen or magnesium. It is rare in the wild and commoner when humans ask too much of a tree in poor soil, especially in urban areas or when we've been busy upending a wilderness.

---

*It's extraordinary the way discovering one feature makes similar ones shine out. For weeks after returning from Scotland, I couldn't help noticing shifts in colors on conifers. I saw the blue tint almost every day, but others that I hadn't looked at for years suddenly leapt out again. There is a healthy golden color that spreads over some sunlit conifers, especially their southern side. It is a pretty genetic effect and commercial growers favor it, which is why you'll often spot it in garden trees.

The yellow color is interesting because it is a negative effect: We see yellow, but we are actually looking at an absence of green. Chlorosis is a sign that the tree lacks the ingredients it needs to make chlorophyll. This is worth knowing as it helps make sense of other leaf-color puzzles. Instead of wondering what created a yellow or orange color, we can often unravel the mystery faster by asking ourselves, where has the green gone and why?

Water, pH levels, and disturbance all have an impact on leaf colors. We will see color changes in any landscapes that range from damp low country to higher, drier areas or from pristine wilderness to busy woods. This is why we will always see the colors fluctuate within the same species as we look down on a forest from a height.

There are times when spikes in the acidity in the soil will paint the leaves strongly. Acidic soil is low in nutrients and no landscape has uniform pH levels. I try to spot this when I'm looking down from mountainsides in areas where there is mining, as this always leads to major fluctuations in water, disturbance, and pH levels. It is usually possible to spot the ripples in colors near the greatest activity.

On the flip side, some trees, especially conifers like Norway spruce, are happier in acidic soils and these trees will lose some of the rich green color in their foliage if the soil is too alkaline. Rivers and roads change water levels and soil chemistry, so it is rare for foliage to hold the same hue all the way to its edges.

There are always a few variables, so the simplest way to think about it is that a uniform rich green color in trees is a sign that the key influences are in the range they can tolerate happily. If we see a patch of trees with an "off" color, it is a sign that one of these variables is nudging danger levels for that species. If we have ticked off water and disturbance as likely causes, problems with the soil chemistry are worth considering.

OBVIOUS AND INVISIBLE

On a walk a few years ago, I was giving the color of leaves as much attention as the slippery chalk underfoot would allow. I enjoyed noticing the darkness in the shade leaves and the loss of color in some of the leaves high up on the windward side of an oak that had suffered during a gale.

After stopping by a field maple near the path, I decided to look at a dozen of its leaves in greater detail and to stay true to my belief that meaning and value will always be found in the colors we see. I struggled to find it. I did not succumb to the belief that it wasn't there, only that I had failed to discover it, so far. A few minutes later I repeated the exercise on a young oak a little further along the same path. The same thing happened: I failed to notice any striking message in the color of those oak leaves either. I won't lie; it was a little frustrating. The colors in each group of leaves were similar to one another and to those of the other tree, too.

Yet I sensed that something was different in the color I was seeing. It was hard to gauge what it was from my memory of the earlier leaves, so I plucked a couple of leaves from the oak and walked back to the maple. When I held the oak leaves next to the maple leaves, the colors were similar. They were also deeply lobed leaves, but there was no confusing them: Their shape was quite different. And there was definitely something different in their hue that was obvious, yet hidden. Was it a stronger gloss on the oak leaves? No, that didn't seem to be it. Then it struck me. Something leapt off the leaves that had remained invisible to my eyes for many years, but now it burst out. The veins were totally different.

The maple leaf had veins that radiated out from the base of the leaf in lines that ran to each lobe, but the oak had a strong central rib with weaker lines running off that highway to each lobe. The vein lines were a paler color than the main leaf and their patterns broke up the color differently. Instantly the colors I saw no longer seemed similar. I would liken this sensation of revelation to when we're looking down on an aerial photo of a large city and it all seems a bit samey. Suddenly we spot a neighborhood we know well, the generality disappears and the particular shines out. And the wonderful thing is that the brain clearly enjoys this sensation and clings to it. Once the pattern has leapt out, it can never hide again.

Knowing that some trees, including maples, have main veins that radiate out from a central base near the stalk, and others, including oaks and beeches, have a central rib, helps explain many of the subtler variations in color we see. In autumn, for example, we may notice shifting color patterns within individual leaves, which are often closely related to the main vein pattern. Instead of seeing apparently random patches of yellows or oranges within a single leaf, we notice that they are arranged equidistantly from the main veins. We have the map that explains the colors.

Leaf veins are unique and a part of each tree's signature. And, like many visual signatures, we come to recognize certain patterns before we can describe them. There are now many tree leaves that I recognize from their vein patterns without any other clues—dogwood is a good example. There have been times when I have seen old, torn, or broken leaves on the ground and known them as dogwood instantly, thanks to the unique "parallel curved" nature of their veins—despite the leaf shape and colors giving no clues, due to wear and tear. You, too, will find there will come a moment when you recognize the vein signature of the leaves you see, often straightaway. And the odd thing is, soon you will come to know this pattern better than the lines on your own palm.

The clearest, boldest patterns are the easiest to befriend, but you will also meet some beguiling quirks along the way. Like the walnut leaf that has veins that charge away from the main rib, heading with apparent determination toward the edge of the leaf only to give up at the last moment and curl away from the edge.

I find it bizarre that I must have looked at thousands of oak and maple leaves and not noticed that simple, clear difference in their vein pattern before. Now it shines out like lightning at dusk. The act of slowing down and looking can make the invisible obvious.

WHITE LINES

Some conifers have leaves with white lines on them, and some don't. Many species of fir—including the Douglas, silver, and grand fir—have two parallel white lines on the underside of their leaves, but most spruces don't. Why are the white lines there?

They are caused by a phenomenon called "stomatal bloom."

Stomata are the small openings that all leaves use for gas exchange. The holes are necessary, but also a weakness. The leaves can't be airtight; they have to exchange gases to photosynthesize, but every opening is a chance for water to escape and this is one of the resources that a tree must guard most carefully. It makes sense that most leaves will have more stomata on their undersides, where exposure to heat and water loss is less severe and space for photosynthesis is not critical.

Stomata are easy to see under a magnifying glass but too small to be spotted easily with the naked eye. However, some species have a white waxy protective coating around the tiny holes. This is stomatal bloom and we see its white lines on the underside of many fir leaves.

It is satisfying to know what the white lines are, but things grow more intriguing when we go deeper into this small mystery. A few species also show white lines on their upper surfaces, but why? The answer is related to the blue-tree compass we met earlier in this chapter. Trees with stomatal bloom on the upper side of their leaves are protecting the stomata there from the drying and damaging effects of solar radiation. It is more common on trees that thrive in direct sunlight. (Shade-tolerant trees have to use the tops of their leaves for photosynthesis, so we don't see stomatal bloom on the tops of their shaded leaves.)

It is most likely that you will spot the white lines on the upper sides of leaves on isolated conifers, growing in open areas or above the main tree line because this is where sun-hungry species do best. For natural navigation, this upper-side white bloom on leaves at the edge of a forest is a clue that we are looking at the south side of that woodland.

You may have spotted a broader pattern by now. If we see interesting colors on leaves, there is always a reason for it, and if the color is silver, blue, or white, it is a clue that the sun lies behind it. And whenever the sun shapes what we see, we can find a compass.

If you feel anything abnormal in a leaf, it is fair to say to that tree, "I feel your pain." Each time a leaf feels thicker, tougher, stickier, hairier, or sharper than we're expecting, we can be confident that it has gone to some effort to combat a challenge in its life. The only question is: What?

If leaves feel tough, it is a sign that they have to endure difficult weather, hot or cold. Laurel, eucalyptus, olive, and holm oak all endure hot, dry seasons in their native areas and their leaves have a coarse, leathery feel. We have seen how big leaves are a liability in cold winters, which is why there aren't many evergreen broadleaf trees, but the few that keep big leaves through winter, like holly, all have tough leaves. Holly leaves survive year-round: They are unusually thick and feel different from almost all other leaves. Not that many people take the time to feel holly leaves because of their spines, which is a sign of a different challenge.

Trees that grow spines on their leaves are defending themselves against grazing animals. This is a dynamic response in many species, including holly, and the more spines we notice on leaves, the harder they are trying to fend off animals. This is why holly leaves near the base of the tree have more spines than those at the top and why holly hedges that are grazed by a gardener's trimming blades are extremely spiny.

Thorns are not the same as spines on leaves: They form from twigs, but both act as a defense against animals. It is ironic that whenever we come across a tree with either thorns or spines it is worth pausing to look for signs of animals. Most spiky trees are much shorter than the main canopy trees—there's no need to defend against a grazing deer 100 feet (30 m) above the ground. Some animals, including many smaller birds, have worked out that thorns and spines stop fast predators from zooming in and out of these trees, so they make a good refuge and home. At certain times of year, it is especially noticeable: It's rare in winter or spring for me to pass a holly, hawthorn or blackthorn bush without noticing some sign of small-animal activity, even if it is only a songbird flitting between the spines and thorns. Later in the year the birds reciprocate by feeding on the fruits of the same trees, then spreading their seeds. I like to think of it in this way: "Thorns and spines are fingers that point to animals."

Some of the lower plants use individual hairs to hold chemicals for defense, like the acid-tipped hairs on the familiar stinging nettle. But hairs on many tree leaves are friendlier and any that feel extraordinarily soft have a downy layer of short hairs. These tiny hairs trap a thin layer of air next to the leaf, which protects it from evapotranspiration—they stop the leaf drying out. The boundary layer of air can also ward off frost, and in some leaves the hairs protect against insect attack. They always have a job to do. In many species, including the beech in my local woods, this downy layer is obvious when the leaves are young, but fades as they mature.

Large leaves that look shiny and feel waxy are wearing sunscreen and a raincoat at the same time. The water-proof waxy layer protects the leaf against strong sun and heavy rain and is very common in tropical rain forests. These leaves often have a marked tip at the end of the leaf, and the waxy surface channels the heavy rain down to the tips and off the leaf as quickly as possible. As a general rule, the more pointy the tip of a leaf, the more rain you should expect in the area.

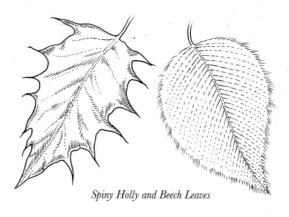

Spiny Holly and Beech Leaves

When looking and feeling for these differences, you'll notice that many leaves have top and bottom surfaces that not only look but also feel very different. The upper side is waxier because it needs more protection from harmful rays, but the lower side is often hairier as it needs better protection from drying out. A downy felt of fine hairs gives white poplar leaves their very bright underside.

My final tip when it comes to feeling leaves is to let your feet join in. I'm not suggesting you have to go barefoot, although there are pleasures in that for sure. Whenever I feel my foot slip on leaves, in town or country, I look for the sycamore and it's usually there: Sycamore leaves decay into a slippery, slimy layer. In woods, a sudden deep spring in the soil is a reminder to look up at the larches that have shed their needles. On the best, slowest days, when your eyes are wandering the branches above, you might feel cones underfoot. Spruce cones are quiet and squidgy, pine cones much crunchier.

STALKING LEAVES

Have you ever walked past a street-food vendor and felt perfectly capable of resisting the temptation, only to catch a whiff on the breeze and fold? I'm immune to such scenarios, but I've heard that the effect can be very powerful around the crêperies of Southern Brittany, France.

Fruit trees are in tough competition for attention. They need to attract insects for pollination, but they're up against every other insect-pollinated plant, and time is short. Flowers act as attractive signposts, but they don't always close the deal. That is why many fruit and some nut trees have "nectaries," swellings at the base of their leaves that secrete sweet energy-rich nectar that insects find irresistible.

I like to feel these jutting protuberances near the base of the leaves of cherry, plum, almond, and peach trees. And when I do, I think of the bees trying to fly past the trees' flowers only for their willpower to crumble as they swerve toward the crêperie—I mean nectary.

While you're feeling around near the base of the leaves, you may notice that each leaf stalk has its own character, too. The colors vary—some have a distinctive red, especially when young—and the shapes vary more than we might imagine. Most stalks are roughly round in cross-section, but any deviation from that is worth some thought and inquiry. Any stalks that are flat instead of round have evolved to allow the leaf more flexibility.

All leaves move in the breeze, but their flexibility varies a lot with species. Trees that grow in bright open areas have leaves that wobble in the wind. Most pioneer trees' leaves, like birch, are fast-fluttering, and those of many shade-tolerant trees, like laurel, are steadier. We can look for the effect in conifers, too: Pines and larches love sunlight and their needles flow a little with the wind. Yews and hemlocks cope with deep shade and their foliage only gets going in a gale.

The light-loving poplar family has super-mobile leaves, and in the case of the aspen, the effect is so striking that it has come to define the tree. Nature writers sign up to courses where they are forced to write "quaking aspen," "trembling aspen," and "fluttering aspen" one hundred times on a digital blackboard. They are then locked into a room with *Roget's Thesaurus* and told they will not be released until they have worked out how to avoid these clichés. They emerge to write about anxious, agitated, and neurotic aspens, furthering nothing in the known world. I digress. Trees with foliage that flops about in the slightest breeze have worked out a way to allow all the leaves on the tree to bear the wind and share the light. And that is why they have flat stalks.

Have you noticed that although builders use a lot of steel girders, they are rarely round or solid in cross-section? You will see H, I, L, T, and U shapes in steel girders, but not round ones. The reason is not that round girders are weak: It's because they are very heavy for their strength.

Engineers know that some shapes give you all the strength you need with none of the weight you don't. Whatever you are building, you don't need strength in all directions. Weight is a dependable downward force and always will be, so engineers don't lose sleep worrying about what will happen to things like bridges if gravity suddenly reverses direction. Every girder shape can be chosen for the forces it will have to deal with, not the ones that will never trouble it. And nature worked out the same thing a little earlier, more than a hundred million years earlier, in fact.

Sometimes you will spot a leaf stalk that is U-shaped. This is a sign that the plant wants to hold something heavy without all of the weight of a round stalk. You will see the effect in varying degrees in lots of plants and trees, but the bigger the leaf, the more likely you are to notice it. If you have ever seen a frond that has fallen from a palm tree, you may have spotted that it has a distinctive U or V shape in the stalk. The palm is on the ground because it experienced an unusual force, like a gust of wind from a direction that the shape was weak at resisting. You will spot this effect in smaller plants, too, including rhubarb.

ON MANEUVERS

Leaves orient to capture the sun's light, and it's not just broadleaf plants that do this: Conifers do it, too. As early as you can make it, ideally on a day of fine weather and just before the sun rises, choose a plant that is out of the wind and study the leaf. Look where the leaf is facing and pick something on that line.

Now repeat the experiment before sunset and you'll spot a difference. There is something of an art in picking your leaf: You need one that will flex during the day, of course, but not one that will be twisted with the slightest breeze. I tend to pick a few leaves on the same plant and average out the direction in which they're facing. I'd recommend starting with the broad leaves of ground plants, then moving to broadleaf trees and finally conifers.

Many plant leaves respond to changes in temperature, hot and cold. During heatwaves, plants lose more moisture through their leaves than they can replace, which weakens the water pressure within the plant. Since that was what held up the leaves, they droop. We see them wilt.

The rhododendron is a diminutive tree, rarely growing above 15 feet (5 m), and many would argue it is more a shrub than a tree. It has some fans and many enemies because of its invasive habits; once it gets a hold in an area, it spreads with determination and can barge native species out of its way.

There are lots of different rhododendron species, but most prefer acidic soil.*

We have recently had one of the mildest Decembers on record in the UK. But Jack Frost will be back before long, and when he is, I will drive north for half an hour to a place called Black Down. It is the highest land in the county of West Sussex, in South East England, and a great place to explore after snow. Black Down is no giant: It is less than 1,000 feet (300 m) high, but that is enough to make a big difference to snow levels in my part of the world. (The effect is compounded because I'm heading inland, and snow gets deeper when we move away from the coast.)

Black Down sits on a geological feature called the Greensand Ridge, a line of acidic sandstone rocks that have resisted the elements better than the rocks to the north and south. Before I even step out of the Land Rover,

---

*Literally the "rose tree," rhodo = rose, dendron = tree. One Sussex botanical society described it as "beastly" in a newsletter.

I can sense how the acid in the soil has changed all the plant and animal life around me. Conifers dominate; gorse and heather do well.

After parking, I walk uphill toward the summit, then explore the area, lifting my feet above the deepest snow. The rhododendrons greet me; their leaves point at the ground, saying, "It's cold on this spot." Rhododendron leaves are famous for their habit of curling and drooping, pointing to the ground during cold weather.

THE CONIFER NET

When you see a splay of conifer leaves, take a look at what's inside it. Conifer foliage acts as a net, catching many items of interest. Depending on the season and recent weather, you might spot dead leaves, feathers, feces, dust, insects, spiders' silk, pollen, and much more. There have been many times when I have been reminded to look for a bird's nest by a feather lying on a dark green bed.

Looking in this net is a cunning habit to develop, because you will soon spot something else. It's impossible not to notice how the shape of each tree varies with terrain. You'll find it a lot easier to look down into the foliage of yews in woods than pines in open country, for example, because, as we know, the shape of the trees we meet reflects the landscape: Pines like open sunny spots and have few low branches; yews thrive in shade and have many.

10

Bark Signs

I HAVE BEEN FORTUNATE ENOUGH to view some of Vincent van Gogh's paintings in galleries, but there's one I've never seen close up and would love to: *The Large Plane Trees*. It's a dazzling image without any special interest in trees, but one part of it strikes me as particularly interesting when I'm thinking about bark.

It's not one painting, but two. Van Gogh first painted the scene in 1889 and called it *The Large Plane Trees*, but then he repeated the painting, calling it *The Road Menders at St Rémy*. The outlines are near identical. The same trees, figures and buildings appear in the same places. But the second painting is far from a copy and the main differences lie in the colors.

Van Gogh is renowned for his pioneering use of vibrant colors and he turns up the dial in the second picture. He doesn't do it uniformly: The figures lose most of their color, in one case turning into the silhouette of a woman carrying what looks like a basket. But the trees are now super-charged with color. The second version has bold golden yellows in the autumn leaves, but my eye finds the bark more striking: It leaps out.

Plane trees have remarkable bark, described by one observer as "military camouflage" and another as "reversed leopard," thanks to its unusual pattern of colors. Van Gogh was abnormally perceptive when it came to color, but even he once overlooked the show that plane tree bark offers. It is challenging to notice bark, but we have an advantage over others because we are looking for meaning. The skin on two trees is never identical.

THIN, THICK, ROUGH, OR SMOOTH

Brown, grey, olive, rust, red, white, silver, black, smooth, papery, rough, striped, stringy, puckered, spiraling, flaking, bleeding . . . Tree bark comes in many colors and textures. Where to start? With some of the biggest differences and boldest signs.

The contrast is greatest, of course, between different families. Look at the smooth thin skin of a beech and compare it with the rough, gnarled bark of a mature silver birch. Both trees have the same goal, to lift some leaves to the light, so why such differences in their bark? It comes back to niche specialization: Beeches expect to be in dense woodland stands with hundreds of other beeches, a shady well-protected world. Birches have to be prepared to stand alone: They must be ready to brave not only the elements but animals, too. Trees that have evolved to cope with shade have thinner bark; pioneer trees and others that grow in isolation or small clumps, like some fruit trees, tend to have thicker bark. I can see a wild cherry tree from my cabin and I admire its tough bark. It stands at the edge of woodland and its coarse skin is armor against the sun, wind, rain, hail, and snow that will hit it over the year.

(There is an overlapping clue: Smooth bark is a sign that the tree grew slowly, taking the time to fill in the gaps in its bark as the girth of the trunk swelled. Rough bark is a sign that the tree grew rapidly, bursting through its own skin. As we have seen, only shade-adapted trees can afford to grow slowly, the tortoises, like beeches.)

How can we tell the thickness of bark just by looking at it? It's easy if there are any ruptures or major injuries in it, but there is also a way to gauge the thickness just by looking at the skin on a healthy trunk. Texture is a good indicator. Roughness is a sign of thickness; smooth bark is normally very thin. It's not a perfect method, but it works in most cases.

If you're walking in the open, not in woodland, you're much more likely to see rough bark. In my local area, I'm either in the woods and among the smooth skins of beeches, hornbeams, and holly, or I'm passing the rough, coarse textures of willows, poplars, hawthorns, blackthorns, birches, larches, and elders. There are exceptions to every rule: Yew thrives in shade and has bark that makes an anxious grandfather's forehead look smooth; perhaps that's because it plans to be around a few hundred years longer than its neighbors.

The same rule applies to conifers, but it's relative because they all tend toward a little roughness in their bark. Pines are sun-seekers and have very rough bark; deeper in the woods you may find spruces with less coarse bark.

Some trees have very thin bark because it allows them to harvest a little of the light that reaches them. If you spot a hint of green where there is a scratch in the thin bark of trees, especially young ones, you are looking at bark that is happy to do its bit to help the leaves. This is common in young ashes. I like to think of this scenario as evolution's test of teamwork. You know those scenarios where two teams compete to win something: The team that squabbles fails while the team that works together prevails. I imagine many millions of years ago two species of tree were competing and struggling to survive in a shady landscape. The leaves of both young trees turn to the bark and say, "Mate, will you lend us a hand and photosynthesize a bit? It's only for a few seasons. Once we've grown taller, you can get back to your main job of protecting the trunk and branches."

The bark on one tree has genes that say, "That's more than my job's worth." That tree species dies out.

The bark on the other tree says, "No problem. Let's keep the armor off for a few years, soak up some sun and take our chances with the elements and creatures. There's no point protecting the trunk if we starve to death." This is the species we see in shady spots to this day.

At the other end of the scale, some species grow very thick bark to protect themselves against forest fires. Trees, like the cork oak, that have evolved to survive one of nature's fiercest attacks need proper protection. But the message is the same in each case: Thicker bark means the tree is seeking better protection against the elements— sun, wind, or fire.

DRESSED FOR THE JOB

There are hundreds of bark shades and colors, but most tend toward brown, with hints of grey, green, or black. We needn't focus our attention on every shade: We need only question the outliers. If bark has a color that stands out or breaks the typical patterns, it is worth pausing to ask a question that historians might ask, "What problem does this rebellion seek to solve?"

Silver-birch bark is bright white, which reflects light well and protects the tree against the sun's radiation. It's a good solution to a problem faced by pioneers.

In his second version of the painting, Van Gogh picked up the mottled mosaic effect in the bark of plane trees. Planes have a habit of shedding bark in large plates, which allows them to tolerate pollution better than most species and explains why we find these trees in cities across the world. Pollution is a recent problem in evolutionary terms, or perhaps planes were the first to thrive around the fires of our distant ancestors.

Red or purple bark, especially if it's shiny, is a sign of new growth, which leads us to the question of time.

BARK TIME

Find a tall old tree and a small young one. Now compare the bark on the two. There's quite a difference, which we'd expect. Now compare the bark at different levels on each tree. Look at the bark by your feet and the bark at head height. There's a bigger difference than many might guess.

Bark changes as the tree grows older. We know that the lowest part of the tree is the oldest and the bark near the base looks older. Some trees have bark that ages gracefully: The bark of a hundred-year-old beech looks similar to that of one that is quarter its age. But most trees have bark that exaggerates its own character. If there are rough patches on a young tree's bark they will get rougher over time; if there are fissures, they will grow deeper.

Losing and replacing skin is a challenge that many plants and animals face. You need to replace skin, but you can't live without it: What's a creature to do? You could regularly shed a thin outer layer, like snakes and humans, knowing you've got more layers underneath. Or you could shed it in patches, which is what many trees do.

One technique trees use is to hold on to some of the outer layer, even as the inner parts grow and expand, which leads to the mixed patterns we see. When you see a crisscross pattern on bark, you will notice that this is either formed of a series of raised diamond shapes surrounded by recessed valleys—or lower diamond shapes with raised areas between. In each case the lower areas are the gaps formed as the inner layer grows under the older outer layer, forcing it apart. Each species has its own signature as new layers appear. Norway spruce bark looks like drying mud; pines have big thick plates; hornbeam bursts out of a too-small jacket.

All trees shed and replace bark as they age, and it's common for them to lose more higher up the tree than low down—another reason why the base of many trees has so much character. They each have their own quirks. Mature Scotch pines lose lots more bark near the top of the tree than the bottom, which reveals an orange tint to their upper parts. It is most dramatic in late summer—the trees lose more bark at this time of year and the sun is lower in the sky than in midsummer, which accentuates the effect.

The London plane loses more bark plates on its southern side, which makes the north and south sides of these trees look quite different. I've seen this hundreds of times—it's one of the most fun urban natural navigation methods, but I'm still not certain of the scientific reason for it. The sun is the most likely cause. Or perhaps the tree is trying to photosynthesize using its inner bark, or because of sunburn or because frost-thaw cycles are more brutal on the south side. Or maybe because algae and lichen bloom on the south side is clogging the bark. Or a combination of all of these factors. Whatever the cause, it's worth looking out for, because now you know how to use a city tree's bark as a compass, it's hard to resist.

THE BIG CHANGE

Each summer the woodland floor is sprinkled with little green saplings. Ashes are the most common in my local woods. I could touch a different one with each hand and foot at the same time, but I don't do that because it would confirm the suspicion that I'm a bit odd.

A sapling that pokes above the ground will be green and soft; its skin will look nothing like that of a tall thick-trunked veteran nearby. We expect to see significant change in a tree's skin over time, but most people think this is a gradual process. While there are gradual changes over the course of the tree's life, there is also one massive change in the bark.

When a tree is very young, it has a soft skin called the epidermis. At a certain point, which varies with each species, the epidermis is replaced by the tougher, thicker periderm. The periderm has living cells on the inside and dead cells on the outside, a little like our own skin. In many species the tree enhances the defensive strength of the periderm by filling the gaps with tannin, resin, or gum.

It's easy to sense this big change in the bark. Scratching the outer edge of a mighty old oak with a fingernail won't hurt it, but doing the same thing to a tree that is only our height would feel like a wound—and it would be. At this early stage the tree is especially vulnerable because the skin is so thin, but also because the tree transports vital nutrients in a layer near the outer edge of the tree. If an animal or human removes a complete ring of young bark, it severs this vital supply channel, which will kill the tree above that line, a process known as ringbarking, or girdling. Squirrels, deer, beavers, or metal blades make short work of the epidermis.

The secondary skin, the periderm, grows under the young bark. The periderm then replaces the epidermis, but the way it does it explains a lot of the variability we see. There are four main types of tree bark: thin, lined, patterned, and patchy. And the way the big change happens explains each of these. Let's look at the simplest first.

Thin bark. Some trees, including many citrus fruits, holly, and eucalyptus species, have a young bark, the epidermis, that survives into maturity. These trees have a thin and vulnerable skin. Eucalyptuses are notorious for their peeling flaky bark, but lemon and lime trees wear it much more tightly, and most trees in this group appear smooth. In each case, the bark is noticeably thin. It is usually lighter in color than most other barks.

Lined bark. You will have noticed that many trees have long vertical lines in their bark. These include junipers and the "thujas," like western red cedar and northern white cedar. These trees have a periderm that forms in an entire ring, a complete circle around the tree.

Patterned bark. This broad term covers the very many barks that have a rough texture and some order, but it's never perfectly neat. It includes pine and oak trees. Typically, there are slightly raised sections of bark, smaller than your palm in each case. In these trees the periderm is formed in curved lumps.

Patchy bark. Plane trees form a periderm in the same way as the patterned bark above, but the lumps or plates are so big that it creates a distinctive effect, hence the "camouflage" appearance.

LATE CHANGE

In truth, once you get down to the precise mechanics of how each individual species makes this change, it quickly becomes fiendishly complex and technical, and I haven't found it adds much to my reading of bark. It is reassuring to know that there is a simple reason for the patterns we see and that we can delve deeper into an individual tree's periderm story if we choose to.

One more part of the big change is worth noting, and that is the timing. Most trees have made the change by the time they reach ten years old, but the exceptions are interesting and worth looking for.

Wild cherry trees fascinate me and many others who treasure wilderness knowledge. The bark has been used in traditional medicine for centuries, supposedly able to cure anything from coughs to gout and arthritis. When wounded, the bark of all cherry and plum trees exudes a thick gum, which is highly chewy and nutritious. The eighteenth-century Swedish traveler and naturalist Fredrik Hasselqvist tells a slightly suspicious story of a hundred men surviving a siege for two months by feeding only on the gum from cherry trees.

We can recognize most wild cherry trees instantly from their bark: It has a dark, ruddy color with thin horizontal stripes, "lenticels," that allow gas exchange. Many trees have lenticels: They are common in bark and fruit—those tiny brown specks you see on an apple are lenticels—but the lenticels in wild cherry trees are bold and distinctive. (Here's how to remember a natural navigation technique: Think of the horizontal lenticel lines in cherry bark as the rails on a fence that holds the trees in the woods. Wild cherry trees are found at the edges of woodland.)

However, even after many years of fond familiarity with this species, a wild cherry tree I see daily has bark that confused me until I discovered its secret. Some of this tree's bark has the familiar lenticel stripes, but large sections don't and look much rougher. I thought it had a disease. I now know that the mixture of textures is the periderm replacing the epidermis, but it takes place so much later in life than in most other species that I would never have guessed without discovering this habit of cherry trees.

A slightly odd group of trees shares this trait of a late change from epidermis to periderm bark. It includes cherry, birch, fir, spruce, plum, apricot, nectarine, and almond. They may not make the change for fifty years or more, and we must expect the bark on young and old trees to look totally different.

My cherry tree has large patches of bark that have made the big change and patches that still retain the original thinner striped epidermis. In a decade the periderm will dominate and the whole tree will appear rougher.

A STRESS MAP

Have you ever glued two things together and soon afterward felt the dried glue on your hand crinkling, cracking, and peeling off as you move your fingers? I find it weirdly satisfying. Whenever a thin layer lies fixed on top of something else, it will tell a story about any movement beneath it.

If there is any unusual movement or stress in a tree's structure, we will find it written in the bark. Professor

Claus Mattheck, whom we met in The Missing Branches chapter, describes bark as the "stress-locating lacquer of the tree"—its cracks and patterns reveal the deeper stresses that the tree is trying to manage.

Whenever you see a tree that has been knocked off vertical, it's worth pausing to study the bark. If we lean our head and shoulders to the right, the skin on the right side of our torso bunches up and that on the left is stretched. Trees have the same experience: If the trunk is bent over by strong winds, the bark bunches up on the downwind side and stretches or breaks apart on the windward. This leads to bigger gaps in the bark on the upper side and a bunching, crumpling appearance on the lower. The effect is greatest in thick bark; in thin-barked trees, you're more likely to see the crumpling than the stretching.

Trees are adjusting all the time to new stresses, even when nothing dramatic has happened. With a little practice we can start to see how the bark patterns reveal the strains in each tree. A great place to look is at the junction where a large low branch meets the trunk at the "branch collar."

Remember that trees have no plan for the size of their branches. They all start small and light and many are shed at that size, so the tree doesn't expect to have to support a massive limb. If a branch survives into maturity and grows long and mighty, the tree may struggle to cope with the weight and it will sag. The angle of the branch will change, but before that there will be clues in the bark at the branch collar. On the lower side, there will be bunching, and on the upper side the bark may show cracks or gaps. Again, the thicker the bark, the more dramatic the effect.

If you notice that the junction has swollen and an unusually large "branch collar" encircles the point where the branch meets the trunk, this may be a sign that the tree is preparing to sever that branch. It's getting ready to close the gate once the branch drops, to stop any pathogens sneaking in. There is a big difference between a large branch that snaps because it fails in a storm and one that the tree drops deliberately. The fattening of the collar is the clue that it is deliberate.

THE BRANCH BARK RIDGE

As soon as you start looking for these effects, you will notice an interesting line that runs over the top of the branch junction. It's a little like a dark scar in many trees and it's called the "branch bark ridge." To my eyes it also looks like a welding joint, a fair analogy as this is where there is acute tension and the tree is trying to hold the branch to the trunk.

The line exists in healthy trees because the tree has to form a special sort of wood to join the branch to the trunk and support it, but the ridge will widen or show cracks if the tree is struggling to hold the branch up. (If we take a moment to think back to The Missing Branches, you may recall the southern eyes, the small oval patterns left in bark where old branches have fallen off. The eyes often have "eyebrows," dark lines that arch over the eye shapes. These lines are the remains of the branch bark ridge.)

If you don't see the branch bark ridge line, but instead, it looks like there isn't a proper fusion at all, you may have spotted a "bark-to-bark" joint (its formal name is "bark inclusion"), a serious structural weakness in the tree. It looks as if the bark of the branch and the trunk are touching each other but not joined as one. It is much more common in branches that are close to vertical than in those that are nearer horizontal.

If you stretch your thumb away from your index finger, and look at the skin between them, there is no mistaking that your thumb is attached to your hand, a web of skin forming a joint. But now place your palms against each other—the prayer pose—and push the bases of your two thumbs together, so the pads are pressed tightly. In this scenario, one thumb is the trunk and the other is a large branch, our skin is the bark. They are temporarily joined together, but if you look at the junction, you'll see that it's just a dark thin crack that runs down between the two pads. This is a "bark-to-bark" joint and we can see how weak it is: As soon as we release the pressure, the thumbs go their own way.

Bark-to-bark joints form when the tree doesn't sense a branch growing, getting steadily heavier, so doesn't provide the junction wood to support it. Why might that happen? A common reason is "bracing."

If a branch touches another branch higher up—of the same tree or a different one—the second branch can act as a support. The tree doesn't sense that its branch is getting bigger or heavier because another tree is doing the heavy lifting. The tree doesn't grow the wood needed to support its own branch, which leads to a bark-to-bark joint.

If for any reason the second supporting branch breaks away, the joint won't be strong enough to take the strain and it's likely that it will fail. This doesn't always happen instantly. Like all architectural weaknesses it can sit there until a moment of great stress, like a storm, and then it breaks.

This is all about height, scale, and time. It doesn't matter much if a finger-thick branch rests on another for a couple of days—it happens in the hazels near me all the time. But if a small branch develops into a major one and a brace is supporting it higher up, there can be an issue, even if it doesn't cause any drama for many years.

A PROBLEM WITH FORKS

The biggest problems in trees start as tiny ones. If a small branch junction has a "bark-to-bark" weakness and grows to become a mighty branch we have a serious problem. The tree has a serious architectural weakness in its structure and no way to wind the clock back. A dangerous failure is likely—it's just a question of time. This is a common problem in trunk forks.

With a single healthy trunk, the stresses are simple. But as soon as there is a fork, we have an issue with gravity: It's impossible for both trunks to hold a vertical line. For a few years the young shoots may grow together vertically, but eventually one, or both, will have to start growing away from the other. This divergence leads to massive stresses at the junction in the fork. If the tree senses the stress early enough, it will form junction wood, and we will see some swelling and a branch bark ridge forming at the dividing line. If it doesn't, it's more likely that a much weaker bark-to-bark joint will form.

A young tree with a bark-to-bark fork may be fine, but it will eventually reach the point where the two trunks are massive and growing apart. Then the stresses at the junction are unsustainable. This is a potentially lethal situation, because half of a large tree can come down. That is why professionals will never let this situation develop for long in parks or other public areas.

The good news is that tree-readers can spot the problem and predict trouble, often decades before it is dangerous. You will probably pass one of these trees over the coming days and be able to predict a terrifying sound and major disaster for that tree sometime in the future. During storms, I think of some of the most precarious forks I know

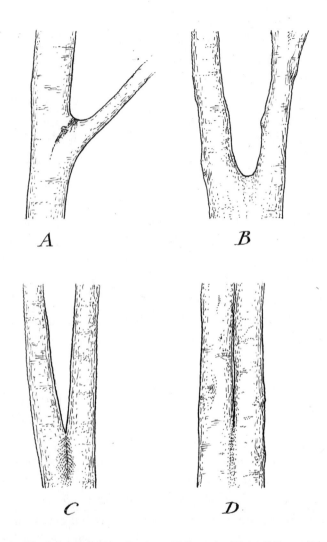

A—The Branch Bark Ridge, B—Strong U-Shape, C—Weaker V-Shape with Healthy Chevrons, D—Bark-to-Bark Joint

well, and about once a year, after the storm has passed, I find one that has come crashing down.

Bark-to-bark joints are the enemy of tree forks, but there is an art to reading the health of stronger joints, too. A gently curving U-shaped joint between two trunks is stronger than a sharp V-shaped one. I think of it like this: If you could trap your hand by karate-chopping the junction, it would be weaker than if you could rest your palm face down in the valley.

If we study the bark, we'll find clues even within the weaker V-shaped junctions. The bark ridge at forks is like an exaggerated version of the same lines at branch junctions; the mechanics are the same but the stresses are much greater. This leads to particular patterns within the bark at the joint. If you see chevrons along the ridge line, try to note which way they are pointing. Chevrons are like arrows: If the arrows point downward, the joint is weak; if they point upward, the joint is slightly stronger. A downward arrow means that part of the tree is vulnerable: It will probably fracture at some point and fall to the ground. If the arrows point at the sky, it will stay up for longer. I remember it like this: "If the arrows point at the ground, that is the direction things are heading."

We have seen how two branches can fuse together where they touch each other. If you ever spot this in two large branches, above the fork in the trunk, for example, look at the fork very carefully. You are likely to spot a bark-to-bark joint or other clues that the tree has not developed a strong junction. This is a tree that is likely to fail in the future.

We can look at a tree from many angles, which is surprisingly easy to forget. Once you've looked at a fork junction from the vantage that allows you to see through the gap, move round so that you are looking side-on, from a position where you can't see the two prongs of the fork. Now look at the profile of the tree. Does the trunk swell at the junction? Chances are that it does, and the amount can tell you how much stress the tree is trying to manage. The size of any swelling is our gauge: The bigger the bulge, the bigger the strain.

If we bring together all of these bark signs, we now have many of the tools used by the pros to predict the current health and any future dangers at branch or trunk junctions.

WOUNDWOOD
Whenever a tree suffers an injury that pierces the protection of the bark, a race begins. Can the tree grow a protective layer quickly enough to put a plaster on the wound and seal out the oxygen, or will an invader get in and feast on the vulnerable tissues inside? If fungi or bacteria gain a stronghold, the tree will struggle to cover the wound and very often we will see a change in color or even dripping from the wound. This is a "canker," a broad term for an infection.

Each tree species has its associated fungal, viral, and bacterial pathogens, many specializing in only one species. Cytospora is less fussy than most and particularly successful as a result: It can be found on willows, poplars, pines, and spruces. In my local woods I regularly see it as a white waxy substance, dripping from a wound in spruce trees.

Cankers are like an infected human wound. There is an opening, an infection, "pus," and sometimes strong smells, all leading to a scar. That's not a very savory analogy, but memorable.

If a tree succeeds in covering the wound with a fresh layer, the new growth is known as "woundwood." It creeps in from the edges of an injury, like hard, slow treacle. Woundwood is not the same as the bark or the layers underneath, which is why wounded trees have scars that last for many decades.

I recently flew out to Texas. I had been asked to provide some training for an aerial firefighting company, aptly named Dauntless Air. The incredible team of pilots and crew put out forest fires by flying small aircraft down to lakes, scooping up water, then flying on to the fire and dumping the water on the hottest part. I joined one of the lead pilots on a training run to look for signs in the lake's water that I could help them interpret. It was an experience I will never forget. Imagine a cross between a roller-coaster and a log flume ride, only there are no rails and it's ten times louder. And it vibrates so much that your eyeballs feel loose.

It was during this brief trip to Texas that I discovered the claustrophobic roots in the middle of the highway we met earlier. Also, I squeezed in a visit to the Fort Worth Botanic Garden and there met professional horticulturist, expat, and fellow Briton Stephen Haydon. Stephen introduced me to some cherry trees with serious vertical scars down one side only, their southwestern side. I could clearly see the woundwood inching in at the edges. It was obvious these trees had suffered some drastic injury and the fact that it appeared on only one side was fascinating—any natural features that have a preference for a certain aspect are catnip to a natural navigator. I asked Stephen what had caused them.

"That's sunscald injury. We only get it on the southwestern side of the cherry trees."

I was bouncing up and down with joy. This is a phenomenon I've been aware of for decades. It's notorious for appearing between the south and west side of trees and is also known as Southwest Winter Injury. I thought it could make a rare but wonderful compass. After years of searching, I'd struggled to find clear examples, yet here they were, in a botanic garden in the Lone Star State.

Sunlight bouncing off frost on a cold morning makes for a beautiful scene, and it's hard to imagine the forces that are at work. The expansion during a freeze-thaw cycle is literally a rock-breaker and can cause serious injury to any plant. Texas is known for its dry heat, but it also experiences massive temperature swings. Even in the short time I was there in March, a warm southerly wind switched to a northerly overnight and the temperature dropped almost twenty degrees between sunset and sunrise. If there are freezing temperatures overnight and a tree heats rapidly in the sun, this can kill the delicate layer just beneath the outer bark. The afternoon air is much warmer than the morning air and the sun is in the southwest at this time, which is why the injuries and scars appear on that side. The tree grows woundwood over the injury, which was what we were witnessing in the botanic garden. Cherry trees are especially prone to sunscald injuries because they have a dark bark that absorbs sunlight and, as we have seen, they hold onto their thinner epidermis for longer than most other trees.

(There is a type of bark scar that can easily be confused with sunscald injury. When a tree is felled, it may collide with a neighboring tree and damage it on its way down. This leads to a vertical line of damaged and then scarred bark on one side of the affected tree. It is common and worth looking for in working forests. The trick is to look up at the bark whenever you notice that the tracks of forestry vehicles have cut a swathe through the trees.)

LUMPS AND BUMPS

A semi-famous friend once told me a trick she uses so she doesn't worry about how she looks. If we stare into a mirror before we head out, we may see blemishes and imagine that others notice them, too. But my friend has learned to check her reflection in shop windows instead of the mirror. "That vague glimpse with no detail is what other people see. The few who notice us at all."

We need not worry about the vanity of trees, and when we take the time to look properly at their skin, we'll see imperfections. No tree skin is perfect and it would look odd if it was. We'll frequently come across lumps and bumps in tree bark that definitely don't look entirely elegant.

If you see a smooth rounded bump on a trunk, one that looks like the bark is still covering it, you're looking at a sphaeroblast. They vary in size from that of a plum upward; I once saw one that must have weighed as much as a car. Sphaeroblast is the formal name for this outgrowth, but don't be fooled by it into imagining that tree scientists know exactly what's going on underneath. They don't. What we do know is that trees have buds under the bark that can grow when hormones tell them to. And we know that they also grow wood to deal with injury. Sometimes this process goes to plan and sometimes it doesn't. The rounded lumps on the bark are a sign that the plan has gone a bit weird. The good news is that these bumps, even the large ones, are not a sign of a major illness or weakness in the tree. Like warts, they are normally more of a cosmetic issue than a serious health problem.

If you see a rough bump, one with a surface that looks gnarly and very different from the surrounding bark, it's a burl. A burl is a specific type of sphaeroblast. Again, the science is a work in progress, but these rough bumps are normally the result of an injury, virus, or fungus triggering the buds to overreact, leading to a large rough growth on the tree.

Recognizing healthy bark is a lot easier than diagnosing a precise problem, which can be tricky even for experts. It reminds me of the famous Tolstoy line from the start of *Anna Karenina*: "All happy families are alike; each unhappy family is unhappy in its own way." Bark can play host to thousands of different organisms, including mosses, lichens, and other epiphytes. Most of these guests do little harm to the tree, but some are a sign that the tree is in trouble.

Below ground, trees can work in partnership with fungi, but above the soil each tree is vulnerable to the pathogens that target them. As a rule, mosses and lichens do little harm, but fungi that sprout out or cause fluid to drip are more serious.

Fungi fruiting above roots can be part of healthy teamwork, but fungi growing out of bark are a different matter. No tree wants a fungus feeding off its trunk, so if you see some there, then it's a sign that the tree is under attack or already dead. The bracket fungi I see on birches invariably mean that the tree is in trouble or has already died—it's common to see them sprouting out of a trunk that reaches about 20 feet (6 m) off the ground, then stops suddenly. There is also a large family of fungi, the saprotrophs, that grow only on decaying wood and don't bother with healthy trees. Either way, fungi sprouting out of a tree's bark is not a healthy sign.

ANIMAL STORIES

You are very likely to see strips of bark missing in two areas: near the base of the trunk and near the junction where branches meet the main trunk. Animals will feed on bark, especially of young trees, and the scars can last decades. Mice, voles, rabbits, and deer all eat bark. Squirrels not only gnaw at it for food but also to send a territory signal to rivals. If you see bark stripped near a branch junction, squirrels are likely to blame: They will use a branch for purchase, gripping it while feeding off the bark near the junction.

Deer don't climb trees, so their vandalism will be seen near the base, but often a bit higher than we might guess. I remember seeing hundreds of short, vertical lines that ran all the way up to head height on oak trees in the New Forest, the national park on the south-central coast of England. It was the work of hungry fallow deer who rear up and rest their forelegs on the tree to gain extra height.

As we saw in the Branches chapter, if you do spot places where the outer bark has come off, it's worth looking more closely. Focus on the inner bark revealed beneath and you will sometimes see small "pimples," tiny protrusions poking up: These are the dormant buds that lie under the bark waiting to spring into action if a chemical message tells them to. Unfortunately, the loss of bark over them means they probably won't survive, but it's interesting to see them all the same.

CURVES AND TWISTS

Whenever you spot a tree with a curve or a lean, study the surface. Mosses and lichens are very sensitive to moisture levels and these vary enormously whenever a tree's trunk leans or bends. The upper side stays wet for longer after rain: The moisture makes a good home for mosses, and nurtures different lichens to the underside.

A curve or lean is not the same as a spiral. Some trunks look twisted, and you may see a spiral pattern in the bark. There are two main reasons for this: genetic and environmental. Some species, like sweet chestnut, love to twist. A tall specimen stands in front of a good friend's home—it makes me dizzy just looking at its spiraling bark.

If a tree is exposed to twisting forces, for example, if it is hit by winds on one side at the edge of woodland or has just lost a neighboring tree, the branches may twist the trunk around. This effect is easiest to spot in smooth bark.

Trailblazing is an ancient practice and a word with a new meaning. These days if someone is described as trailblazing, they are typically moving fast in new territory. It's a popular metaphor, but a slight perversion of the original meaning. To "blaze a trail" means to mark it so that you can recognize it if you need to retrace your steps or help others to follow. When I walked across central Borneo with Dayak tribesmen, they would slash at trees along the way with their long blades. This would be considered vandalism in many countries, but for the Dayak, it's just a prudent way of marking their route. A similar habit pops up in many folk stories, because it's pragmatic—it's simple, it works, and everyone can follow it, literally.

If you see unnatural marks on trees, bright lines, and splotches of color on bark that are clearly not the work of nature, you are looking at deliberate signs, modern blazes. There are two main reasons why humans mark trees with paint or other bright substances, and one has the same ancient roots we've just looked at. If a running, cycling, or other race event passes through woods, the organizers often use paint marks on trees to blaze the trail. (There's a more environmentally friendly technique that is growing in popularity and it is reminiscent of the "Hansel and Gretel" story. On the mud at path junctions, little arrows of white flour point the way for the competitors to turn. Temporary but better than bread crumbs.)

There is another reason for paint on trees, which is not good news for the tree. Foresters signal to each other using paint and each mark is a sign of what action to take with each tree. For example, a lead forester will scope out a managed woodland for sick or dangerous trees, then mark those trees with a luminous splotch of paint. It is nearly always the mark of death and the next team through will fell them. Sometimes the code is less severe: One color is for felling and another for taking off a dangerous limb. Have fun cracking the cipher.

Finally, there is a bark pattern worth searching for on branches or trunks that have come down. If you see one lying on the ground, see if you can spot where animals cross over it. When an animal passes over a branch or log, they scuff the bark with their feet. Animals are habitual so where it happens once, it is likely to happen a thousand times. You can practice looking for this effect on human paths through woods: Notice how a fallen trunk that hasn't been cleared away will show scuff marks where people and dogs step over it. This soon strips the bark from that part of the tree altogether. Once you've noticed it there, you're ready to spot it on deer trails and other animal highways. The woodland animals are blazing trails over bark that we can follow.

The Hidden Seasons

THE TREES ARE BARE in winter; leaves burst out in spring; there is a full, then fruity feel to summer; and the leaves fall in autumn. Repeat. On to the next chapter. Not so fast.

My family regularly berates me for dressing like I live in a ditch. For almost all my working hours I wear clothes that mean I don't have to hesitate if I want to sit or lie on muddy ground to get closer to the action. Even as I write this in a warm, dry room, I am dressed in filthy outdoor gear, because the urge will take me sooner or later to head out into the woods. But, very occasionally, I have to scrub up.

About five years ago, I was dressed in a pale cream linen suit as I sat down to enjoy lunch with my literary agent and UK publisher on the roof terrace of the publisher's impressive London HQ. We sipped our drinks in the sunshine and looked down at the boats passing on the River Thames far below. On the table next to us, a gang of highly caffeinated young publishers gesticulated with their cup-free hands as they discussed launch plans for a new book. For a few dangerous seconds, I felt more like Byron than a tramp, but I resisted the temptation to lean over the railings and declaim at the passersby.

> There is pleasure in the pathless woods,
> There is rapture on the lonely shore,
> There is society where none intrudes,
> By the deep Sea, and music in its roar:
> I love not Man the less, but Nature more . . .

The three of us had met to catch up and discuss ideas for my next book. After ten minutes of pleasantries, I edged my chair forward and threw out my idea: "Bare, Bud, Burst, Flower, Fruit, Fall—The Six Seasons of Trees." I left it hanging.

My publisher made a face as if I'd turned up in my outdoor gear. "I'm not sure. Once you start breaking up the traditional seasons, can't you just keep splitting the year infinitely? Where does it stop?"

At the time I was surprised and a little disappointed by his reaction, but he had a point. There is a Japanese tradition that each year holds seventy-two micro-seasons.

We left that meeting in good spirits, but with no firm plans. I still like that title, but I like the idea behind it even more.

The idea for that book had been driven by my excitement in seeing the major changes in the trees that are lost when we think only in terms of four seasons. Many will leap out if invited, fewer than seventy-two, but enough to make the idea of four seasons seem naïve. The key to spotting these changes is focusing on the edges of the traditional seasons, and we will start by seeing spring before others do.

SPRING PINK AND PALE

Each spring, I'm on the lookout for a particular seasonal moment. This year it was the best I can recall as the sun was full and the wind was just right. I was walking along a broad track in the woods when small pink objects started to rain down from above, carried on a steady breeze. The sun cut through the gap in the trees and lit the dry colored flakes as they fell.

Trees plan quite far ahead. They have no choice if they want to start growing energetically at the start of spring, because there isn't much energy available early in the year—the temperatures are still low and the sun is nowhere near summer strength. The solution is to save up some of last year's energy and package it in little bundles ready for the year ahead. The bundles are called buds.

Toward the end of the growing season, deciduous trees set buds on their twigs ready for new growth in the coming spring. The buds contain everything needed for new shoots, leaves, or flowers and the stored energy allows them to burst into vigorous growth. They can be thought of as a cross between a seed, a battery, and a plan. They are strongly influenced by the conditions of the previous summer, which is why unusually good displays of flowers or fruit tell us as much about previous seasons as the current one.

The buds of deciduous trees are protected by scales, and many have a pink or ruddy color. Before anyone notices the leaves appearing, the buds have swollen and added color, sprinkling the trees with pinks and reds. Once a week from January onward, look at the colors you see in the bare trees and this will allow you to spot the pink that washes over them before the leaves make their appearance. Get in closer to the twigs and you'll see the individual buds. The form and colors of each species are unique (buds can be used to help identify trees) and some species are redder than others. My local beeches put on a determined pink and there will be trees near you that do, too. Early each spring, before the conversation turns to the trees being in leaf, a dry rain falls from the trees. A mix of pinks, reds, and browns will scatter any sunlight as the leaves burst out of the buds and their scales fall to the ground.

Soon there are leaves on the trees, but a couple more spring colors, which are easy to miss, are well worth looking for. Some of the earliest leaves have a pink or red tint, too. The color is caused by a pigment called anthocyanin that helps protect young leaves against damage from excess direct sunlight. The pink-to-red coloration is commonest on leaves on the south side of trees and other plants, like brambles, that are open to the light. I like to think of it as the plants putting sunscreen on the kids.

Most leaves aren't pink, they're green, of course, but even here there are small surprises. The earliest leaves are lighter in color than the ones we will see in mid-to-late summer. Deciduous leaves tend to start pale and darken as the season progresses, especially on their upper side. Most people miss this because they're only looking for the way leaves turn brown as autumn approaches. That's why I enjoy trying to take a mental picture of the deep green leaves in late August. (It is easy to spot in photos, too, but not quite as satisfying.)

I've heard numerous reasons why leaves are especially pale in early spring, but the most convincing is that this is a vulnerable time for leaves and the trees don't like losing too much chlorophyll to greedy animals. The leaves lack color because the trees don't invest fully in them until they are more mature and better protected.

As winter's grip weakens, keep an eye open for the "pink and pale"; then, as summer passes its peak, look for a darkening in the leaves. Soon you will be seeing the seasons that hide between the big four.

DECIDUOUS OR EVERGREEN?

From a good vantage point, it's easy to see how conifers dominate in some areas and broadleaves in others. But there is another division to look out for: evergreen and deciduous. Seen from above, the Crandon area of Wisconsin is home to a few dominant tree species, including black spruce and larch. Spruces and larches are both conifers, but larch is unusual in that it is a deciduous conifer: It loses its needles each autumn and regrows them in spring.

Shedding leaves each autumn means throwing away a lot of water and minerals, even in brown leaves—trees retrieve only about half of the minerals in the leaves before they fall. Around Crandon, the evergreen spruces outcompete the thirstier larches where the soil is dry and the larches win in the areas where water is plentiful. Moist soils also tend to be more fertile, with more of the nutrients that the trees need. This is one niche example of a broader, simple rule: If we see deciduous trees, broadleaf, or conifer, the soil is kind enough.

A popular way to simplify things is to say that evergreen trees keep their leaves throughout the year and deciduous trees lose theirs in autumn, replacing them in spring. Evergreen and deciduous are useful labels, but it's best to think of them as two boxes that contain a collection of habits. Both are oversimplifications that conceal many interesting individual behaviors.

Let's start by looking at the evergreens. There aren't many leaves that can last more than about five years, even on evergreens, because the cells in the leaves will start to pack up around then. But evergreen trees don't wait five years then dump all their needles in one go; they each have their own way of shedding and replacing leaves and it reflects the places where you find them.

THE CLOTHES-SHEDDERS

If we walk out of a cool room into the hot sun, there is a good chance that we will change our clothing to adapt. A layer comes off and we roll up our sleeves. Some evergreens do something similar: they shed large numbers of leaves in times of stress, like drought.

If you live in a dry area, you will regularly spot whole bare branches. It is tempting to think of them as dead, but come back after a wet season and you'll see they have a healthy foliage—the sleeves are rolled down again. (The trees manage this by having some leaves that are deciduous and some that are evergreen.)

The clothing analogy is all right but not perfect: The trees are reacting to a lack of water rather than to heat. The formal name for this habit is heteroptosis. If we both put a penny in a jar each time we use that word over the next decade, it won't buy us one sleeve of a shirt.

WINTER-THINNERS

Some evergreens shed some of their leaves in winter: They thin their foliage before thickening it again. Holly and the American hornbeam do this. As a general rule, the harsher the winter, the more some evergreen trees will shed their leaves. If one of these species covers a variety of climates, we will find that it has few leaves in the dead of winter at the harsh extremes, but plenty in the milder zones.

This effect will vary over large distances, but thanks to microclimates we may spot it on a much smaller scale, too. A holly bush in a harsh frost pocket may have thinner foliage than one in a warmer spot within sight of it. Some botanists call this habit brevideciduous; I call it winter-thinning.

SEMI-EVERGREEN

In 1762 William Lucombe, a horticulturalist working in Devon, in South West England, noticed that an oak he had grown from an acorn was behaving strangely: It didn't drop its leaves in winter.

A few trees, mostly in tropical areas, are known as either semi-evergreen or semi-deciduous. They shed leaves in a short period but replace them almost as quickly. It's like autumn and winter are compressed into a few days.

The Lucombe oak is a hybrid, closely related to the Turkey oak, and it survives to this day, but in tiny numbers. Apart from that one strange hybrid, a handful of other species, like Brazilian teak, also have this habit, but it's not something any of us are likely to see frequently, so I mention it only for interest.

After growing clones from that original parent tree, William Lucombe chopped it down in 1785 because he wanted to be buried in a coffin made from its timber. He stored the planks under his bed, ready to be used for his final resting box. By the time he died, at the impressive age of 102, the wood had rotted in the damp air of his Devon home.

Winter-green

We expect to see deciduous trees with a full canopy in summer and a bare one in winter. This works well when winter is tough and summer is kind to trees, which is what we find in temperate climates. But in parts of the world where winters are kinder and summers are harsh, trees have learned to flip this rhythm on its head.

In the Mediterranean climates—which are spread around the world and include parts of Chile, South Africa, and California—summers are bone dry and burning hot, but winters are mild with rain. In these regions, trees like the California buckeye (also known as the California horse-chestnut), have full foliage in late winter through to spring, then lose their leaves by midsummer. It is another example of the power of microclimates. The closer to the coast you get in California, the milder and moister the summers and the more likely it is that the tree will hold onto its leaves through the summer.

SMALL MEANS EARLY

If a small tree has evolved to grow under the shade of a full canopy, it's fair to say it must have developed a way of making the best of a bad lot. It can help to take things slowly and steadily. There may not be much light near the forest floor in summer or winter, but over the whole year there is more than enough for a small tree. A simple solution is to be evergreen.

If you walk through deciduous woods in winter, you'll quickly spot that a few smaller tree species still have their leaves. I regularly see holly, yew, box, and others, all in deep shade in summer, but happily harvesting low light levels at other times—they do especially well in early spring and late autumn. (If you walk under the conifers of an evergreen wood, perhaps spruces or firs, you won't see any of these smaller evergreens, which demonstrates the importance of the light they get under deciduous trees in winter, spring, and autumn.)

When you are looking for the pink flush of the buds in late winter, make sure you lower your eyes. Many wild-flowers know the clock is ticking—soon there won't be much light left in woodland at ground level. The early flowers have to make sure they beat the trees to the sun. In my local woods the bluebells beat the beech canopy and put on an extraordinary display, a magical lilac carpet, which people travel from all around to witness.

The smallest trees use the same trick as the wildflowers and come into leaf before the taller canopy trees. In my part of the world, hazel, elder and hawthorn always beat beeches, ashes, and oaks.

The size rule even works within the same species: Young trees come into leaf a couple of weeks before their parents and this is one of my favorite sights in early spring. Each year there are two weeks, normally in mid-April, when I can walk through my local woods and spot wonderful colors. There are still no leaves in the main canopy high above me: If I look vertically upward, the sky is easy to see and I can watch clouds pass between the silhouetted branches. But if I lower my head and look horizontally through the woods, there is a healthy covering of leaves. The youngest trees have stolen a march on the older ones and thrown out leaves to capture some of the early light before it is too late. It may be the only generous helping of direct sunlight they will harvest that year.

As soon as you spot this for yourself, you'll also notice that two effects combine powerfully here. The first leaves on the younger trees are very pale indeed. When the sun reaches through the bare canopy to highlight them it creates a striking scene. There is a fluttering sea of brightly lit pale leaves at head height, but none in the canopy above. Nobody will fail to be moved by this sight, but knowing to look for it improves your chances of meeting it, and understanding why it happens adds a layer to the experience. It is quite wondrous.

WHEN IS THE RIGHT TIME?

One cold, wet January afternoon at home, I lit a fire, made a pot of tea, and settled into a comfortable chair with a copy of an article published by the American Philosophical Society in 1963:

THE SCENT OF TIME
A Study of the Use of Fire and Incense for Time Measurement in Oriental Countries

Thanks to the writings of the poet Yu Chien-ku, we know that incense sticks were used to help keep track of time in sixth century China. And by the T'ang Dynasty (618–907 CE), incense clocks had grown more sophisticated and could be used to monitor how long monks spent in meditation.

Time is an essential part of navigation, and over the years, I have enjoyed learning about many intriguing early timekeeping devices. Long before atomic clocks or iPhones, there were sun clocks, water clocks, and candle clocks. In our home we still burn an Advent candle each December, although we often forget for a few days, then have to burn through them; at this point we forget again and burn ourselves well into the future. This leads to laughs and increases my respect for those who would have been punished for such poor discipline in the last millennium. Perhaps somewhere in your cupboards at home there is one of the many board games that keeps time with a simple hourglass filled with sand. These games are not the most relaxing.

There are many ways in which humans have learned to gauge time, daily and annually, and they each have their strengths and weaknesses—for instance, water clocks run slow when it's cold. Nature has many clocks and calendars and they all work well, but each has its foibles. The basic rule is simple: Astronomical cues are more dependable than the weather, but plants need to be sensitive to both.

We can say with certainty the precise minute that the winter solstice will be, but not whether we will see the sun on that day. We struggle to predict which week a tree will come into leaf and we can't even accurately predict whether it will beat the neighboring tree, even if it has done so for the past five years. All of which prompts the question: How *do* trees know when it's spring?

We have a decent, if not perfect, understanding of how trees measure time. They gauge the season in two main ways: They measure the length of the night and the temperature. As winter turns to spring, the nights shorten and this is the most dependable part of the tree's calendar. If trees gauged only the length of night, spring would occur more like clockwork: We could expect to see leaves appearing on the same day each year. It would be a bit boring, so I'm grateful that is definitely not what we see.

The temperature part is more slippery. Summer will be warmer than winter, but there are surprises in spring each year: A week in April can be colder than one in February and regularly is.

We know that there is an advantage to coming into leaf early, especially for small plants and short trees. But there is a risk, too: Deciduous leaves struggle in sub-zero temperatures and a single frosty night can be fatal. The aim is simple: Come into leaf as early as possible, but try to miss the last of the frosts. Each decade outlier frosts come abnormally late: They kill lots of plants and some trees, so the aim can never be to miss all frosts, because the only way to do that is to miss spring altogether. Plants have to take a calculated gamble—deciduous trees are in the business of risk management.

The length of night is the giant pendulum that allows trees to get to roughly the right part of the year whatever the weather, which is why we don't see trees coming into leaf during weird January heatwaves. Using temperature to gauge time is much trickier. Trees don't have a crystal ball and can't predict the weather. All they can do is monitor what is happening and what has happened. One clever trick they muster: They can count, and each species keeps track of how many warm hours there have been. Sugar maple trees in Canada, for example, need 140 hours of warmth before they will call it spring. And this temperature tally clock runs throughout the seasons:

Flowering times, leaf fall, and dormancy all have their triggers. The same sugar maples have to count off two thousand hours of cold weather before they sense that winter may have passed.

The way many species count warm time is interesting. Trees are sensitive to temperature and duration, so a shorter spell of much warmer weather counts the same as a slightly longer spell of mild weather. This counting method is referred to as "thermal sum," or "degree hours." It's hard to visualize, but we can think of the total warmth needed as being like sand in the hourglass. The tree will not come into leaf until all the sand has run into the bottom half. This may happen steadily, two weeks of mild weather, or run much more quickly, seven days of warm weather. (In this analogy, the hole between the top and bottom of the hourglass opens up to be much bigger during heatwaves.)

Many fruit trees need a chilling period in winter, too, and without it won't flower or fruit. This has always seemed odd to me: It's as if these trees don't completely trust the long nights and need to be absolutely certain that there has been a winter before they can believe it really is spring. As with warm temperatures, the plants count the cold days; some, like sugar maple, need a lot more than others. Winters in the UK are sometimes only just cold enough to convince the beeches, which need a lot of cooling. This leads to late leafing: Mild winters make the trees hesitant in spring, which could make them very vulnerable to climate change.

White and blue colors caused by the plentiful wax on the leaves on the southern side of a spruce.

A hazel leaf with a pronounced tip for channeling off rain. Pointed tips are more common in wet regions.

A cherry tree's bark makes the "big change." It sheds the young epidermis, with its lenticel stripes, to reveal the tougher periderm below.

Walnut trees in Wiltshire, in South West England, have poisoned the soil around their roots.

A weak fork with a "bark-to-bark" joint, with small "southern eyes."

The same tree, seen from the side. Notice the swelling of reaction wood at the fork as the tree contends with the stresses there.

Flowers or fruits at the end of the branch lead to messy structures, as in the wayfaring tree.

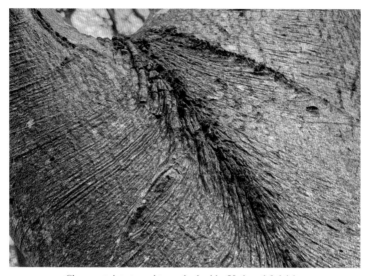

Chevrons point upward toward a healthy U-shaped fork joint.

Stephen Haydon shows me the sunscald injury on the southwest side of a cherry tree in Fort Worth Botanic Garden in Texas.

A giant burl on a beech tree in my local woods.

Fungi protruding from the trunk are a sign of a tree in trouble. After many years watching bracket fungi thrive on this beech, it eventually failed during a storm. There are also epicormic sprouts at the base of the tree, another sign of stress.

The natural spiral in a sweet chestnut's bark. The older, lower branches point down.
Leaves gather in a wind shadow at the base.

The pink case of a maple bud.

Roots adapt to the shape of the land and any stresses on the tree. They are broader and shallower than many expect.

Woundwood creeps in to seal the opening where a major branch once was.

Autumn colors start high up on the southern side of a beech.

Broadleaf trees dominate the valley, and conifers the higher ground, in Snowdonia, Wales. Flagging in the tall conifer poking above the ridgeline. A trunk-shoot compass on the oak on the left. Autumn colors are stronger on the right, southern side of the exposed broadleaf trees. We are looking southeast.

The coldness clock has a powerful effect in species like apple, apricot, peach, and many nut trees: An abnormally warm winter can devastate the crop for farmers the following summer. The whole peach crop failed in the southeastern US in 1931–32, following an unusually mild winter.*

It may seem a bizarre system, but the weather can behave very oddly, too. The trees are trying to recognize all the different ways that spring breaks winter. We may have very cold weather for three weeks broken by a heatwave, or it can be mild for weeks on end. Combining astronomy and weather clocks is the tree's way of trying to beat the frost without waiting too long and missing all that good light. If you've ever planned a big gathering outdoors in April, you will have some sympathy with the challenge the trees face.

At this point we might open our desk drawer and pull out our badge that reads "Evolution's Assistant" and interject, "Wait a minute. Surely it would be a lot easier to stick with the ever-dependable astronomy clock. Let's just wait until the nights are the right length and call that the start of spring."

But try it. Pick a date for a deciduous tree you know well to come into leaf and write it down so you can't cheat. Then watch it over the coming years. What you'll probably

---

*The temperature clock varies across species and subspecies. For plants with a valuable commercial crop, the research and science are amazingly detailed. Peach cultivars, like Mayflower, will not flower unless the buds have been below 45°F (7.2°C) for a thousand hours; others, like Okinawa, are good to go after just a hundred.

find is that you look pretty smart for a few springs, but then there's a long early mild spell and the neighboring trees beat you, coming into leaf a fortnight earlier than normal and stealing all that good light. Then a few years go by and suddenly your date forces your tree to burst into leaf deep in a late hard frost. Game over. Nature will tolerate many things, but rarely total loss of energy and death. It really doesn't like death.

If we continue this thought experiment, we could start picking different spring dates for each species in our neighborhood. But then we realize we also have to pick different dates for each individual tree, depending on where it is growing: a later spring date for the oak in the frost pocket than the one near the warm buildings; an earlier autumn fall for the drought-prone trees on the hill than the ones by the stream. By the time we have done this for all the trees in the area, we could venture out and do it for the county and then—why stop there?—we could do it for the whole country! At about this point, we might feel a little exhausted, having just spent a minute considering the effects of a Douglas fir's shadow on a young birch and whether that should change the date of spring for that one tree. This might be the moment at which we feel grateful that each tree takes care of this for itself, gauging the light and heat in its exact spot. This is why we see seasons roll over countries, spring hitting lower latitudes before higher ones, for example, and why spring already comes earlier to those oaks by the warm buildings. We're lucky: We can put away our clipboards and let the tree clocks do their thing. They may not be perfect, but they know what they're doing.

Each species gives a different weight to solar patterns and temperatures. Small trees rely more on the length of night; temperatures near the ground fluctuate wildly, so even in shade, light is more dependable than temperature. Trees are less sensitive to length of night than smaller plants, but Scotch pine and birches are more attuned to it than most other trees. The unique timing of each tree can be traced back to its character and frailties. Mulberries are famous for their habit of suddenly putting on a heavy cloak of dark juicy fruits. I remember my mother's love-hate relationship with mulberry trees: good for foraging, bad for laundry. Like most plants with soft hanging fruit, they don't tolerate frost at all well—mulberries are one of many fruit trees that take their time coming into leaf in spring.

Folklore about the race between oak and ash trees purports to foretell weather.

> If the oak before the ash, then we'll only have a splash.
> If the ash before the oak, then we'll surely have a soak.

It's nonsense: No plants can predict future weather events. They reflect past and current weather. But it's interesting for other reasons: Oak and ash are relatively late in spring because they have the same weakness. They grow new vessels before coming into leaf that are especially vulnerable to frost. (The fact that they sometimes beat each other is a reflection of the temperature clock. Oak comes into leaf eight days earlier for every degree rise in temperature; ash reacts less enthusiastically and accelerates by only four days. This is why oaks tend to win the race in warm springs, ash in cooler.)

There is genetic variation within each tree species. Trees of the same species are not identical, which influences their responses to temperature and light. If we looked down on a forest of the same tree species, all in theory experiencing exactly the same things, we would still see fluctuations in colors during spring or autumn, which makes for a prettier scene.

AUTUMN QUIRKS

We see people grow steadily older until they're in the "autumn of their lives." As they do, they start to look different, a bit more "weathered," then crumple and eventually die. It is tempting to think that deciduous leaves get older, change appearance, crumple, and die in autumn.

However, leaves change and die as a result of a deliberate, active process that is much closer to euthanasia than a long process of natural aging. Dr. Peter Thomas is Emeritus Reader in Plant Ecology at Keele University. It's a good title, but it doesn't do full justice to his tree-guru status: There can be few on the planet who have done more to advance our understanding of trees. I was once fortunate enough to spend time with Peter, looking at trees in an Oxfordshire woodland, in South East England. He suggests a simple, practical experiment that demonstrates the difference between what many think happens in autumn and what actually does—and it's one we can all try.

In summer we can find a branch on the ground, which has snapped off a tree when it still has its green leaves. Over the following weeks we can watch those leaves turn brown and die; it looks similar to what all the other leaves on that tree will do in autumn. But it's not the same, and we can feel the difference. If we take hold of one of those brown leaves and try to pull it off the twig, it resists us. It is still held quite firmly in place.

Leaves fall after the tree draws back the valuable chemicals into the branches, then seals off the leaf during a process called "abscission." The tree doesn't just shut down supplies of water and food to the leaf and wait for it to give up. This process severs the leaf's attachment to the tree, which is why leaves fall in autumn.

Once you have tried this on the ground, you'll start to notice the same effect on individual branches that have died prematurely but stayed attached to the tree, perhaps after fracturing in a storm. There are brown leaves on that branch that last a suspiciously long time, often staying on the tree into winter, long after the healthy leaves have fallen in autumn.

If we take the time to look for something, we will often notice something else instead. If you are looking out for injured branches that have held their brown leaves into winter, it won't be long before you spot some small deciduous trees and lower branches on taller trees that have rebelled and held onto lots of brown leaves, well into winter.

This has nothing to do with injury: It's the result of a healthy process called "marcescence." It is common in oaks, beeches, hornbeams, and some willows, but most conspicuous on the smaller, younger trees or only the lowest branches of the more mature trees. In my local beechwoods, I see hundreds of branches with brown leaves on them at about head height even in January, but none in the canopy above.

Marcescence in beech makes it a popular choice for hedges, which hold leaves for most of the year: green from spring to autumn, then brown leaves until about February, bare for a month or two, then the cycle starts again.

There is clearly some evolutionary advantage to this habit, but it's not obvious what it might be. One idea is that the dead brown leaves are unpalatable and off-putting to grazing animals, so they give the young trees some protection from grazers. Another idea is that holding onto the leaves is a way of sprinkling their minerals over the roots at the right moment, just before spring growth. I could think of worse PhD aims than trying to solve this small mystery once and for all.

We might expect the timing of autumn to mirror that of spring, but a tree's aims and risks are slightly different then, so the way it reads the clocks changes, too. Trees rely much more strictly on the solar clock—the length of night—in autumn, which means that we can predict the date leaves turn brown more accurately than the date the tree comes into leaf. Part of the reason for the switch in emphasis away from temperature is that trees may get a second chance to grow if they lose leaves to a frost in spring, but they don't get any second chances in autumn. If the trees are caught off guard by a frost in autumn, they can no longer draw back all the precious minerals in the green leaves: They are lost to the frost.

The ground, too, has changed in autumn and the land may be parched. The risks of being too early with autumn are less than those of being late, and stress can accelerate things a little. I am writing this on the last day of July 2022, and it has been one of the driest, warmest Julys on record. Newspapers are reflecting the broader national observation: "Trees are dropping the leaves and fruit is maturing weeks ahead of schedule due to record-breaking temperatures and a lack of water."

Direct sunlight accelerates and enhances many natural processes, including leaf color changes, which is why the south side of trees can look quite different from the north side in autumn. During our walk through the mixed broadleaf trees of the small Oxfordshire wood, Peter explained why the top of the tree often turns before the lower parts. There is friction in the vessels that carry water from the roots to the leaves, and the longer the journey the greater the friction. If the ground is especially dry, the leaves at the top of the tree will struggle, turn, and fall before the lower. Together these effects lead to striking differences; the leaves at the top of the south side of the canopy turn gold, red, or brown, long before lower leaves on the north side.

Evolution is a genius, but it struggles to keep up with urbanization. In cities, trees next to streetlights can confuse artificial light with sunlight. Trees on brightly lit streets don't spot autumn coming and stay in leaf for far too long. Each part of the tree is doing its own timekeeping, which means the effect is localized, the leaves on one side of the tree turn and fall, but on the side nearest the streetlight, they stay green. The first winter frost savages the green leaves. It looks as if the streetlights are damaging the trees directly, causing the leaves and branches to suffer, but it is frost that kills the leaves. The artificial light is ultimately to blame, of course: it has scrambled the tree's clock.

The effect of the streetlight on the tree is localized. This idea invites us to investigate an interesting and important concept more deeply and we will do that by thinking about dogs.

At home at about 5 PM each day we feed our pets: two dogs and two cats. We announce that it is teatime by calling them, then rattling the Tupperware container filled with dry food. The cats ignore this for a couple of minutes, in the way that cats do. It's a power thing. But the dogs come running with urgency, often taking corners too tightly and clipping them.

They have heard the call, paired it with their hunger and the time of day; their brain sends messages via their nervous system to their limbs, leading to a mad dash for dinner. A decision by each animal leads to signals that coordinate the rest of it into a single purpose to get to the food as quickly as possible. We are so used to seeing and experiencing this animalistic central-nervous system way of reacting to the world that it's hard not to assume that all other organisms operate like this. But they don't.

Each leaf, each branch, each flower, each root in a tree is sensing and reacting to its own little world. There is no central nervous system. Scientists in laboratories like to lead two parts of the same plant into very different worlds: One has lots of light and kind temperatures; the other is in the dark and cold. The result in intelligent animals might be psychological problems as the brain tries to reconcile two worlds into one, but for the plant, it leads to two sides that look very different.

This localized response helps each part of the tree welcome the seasons at the right moment. The buds at the top will experience a different microclimate from those nearer the ground, and those at the end of branches experience different temperatures from those nearer the trunk. But the tree "knows" this, and the buds aren't identical: They respond differently depending on where they are. In peach trees, for example, the buds at the top don't need to chill for as long as those on the side branches. This is vital and helps balance things. The trees would struggle if all the buds reacted to temperature in exactly the same way because it varies so much over tiny distances. After clear nights in winter there is a layer of much colder air close to the ground: If the tree didn't factor this in, the chilling effect would mean the lowest branches might sense they were in a different season from those at the top.

The trees do their best, but they can't respond perfectly to all the local temperature nuances caused by microclimates. This means we will see seasonal variations in each tree: The buds, leaves, flowers, or fruit don't come out in perfect synchronicity but arrive in flushes over each tree. For example, you will see leaves appear first on one side and at one height in spring.

Once you have invested the time to look for these hyper-local variations in seasonal changes, you may spot one of the most interesting autumn trends: Some trees change color from the inside and others from the outside. Trees that produce their leaves steadily and favor open ground—including pioneers, like birches—have leaves that turn first on the inside of the tree and this progresses outward. Woodland trees, especially those that bring out leaves in a spring burst, like maples, normally turn from the outside in.

AUTUMN WINDS

Last autumn I was walking on a cold, still morning when I witnessed a pairing of events I'd never noticed before. There were plenty of leaves on the ground and plenty still on the trees. Occasionally one or two brown leaves would rock their way down in front of me. Then I noticed that when a few leaves fell at the same time, it meant something: A flurry of falling leaves coincided with activity above my head.

Wood pigeons taking off and squirrels leaping gave the leaves a push, which sent a collection on their way to the ground. Soon afterward, and ever since, I have learned to look for birds and squirrels in the autumn canopy whenever I spot a gang of leaves falling. It is very satisfying.

For every time an animal causes a leaf to fall, there must be a thousand when the wind does. We know that the tree shuts down the leaf and seals it off, which severs the bond that keeps it attached to the tree. But the tree doesn't jettison the leaf—it doesn't force it to fall: It lets it drop off naturally. The final step in this process is often the wind plucking the leaf off the tree. A gale will snatch one earlier in autumn and a breeze will take one later in the season. And this is where we can look for patterns.

The side of the tree that is hit by strong winds loses its leaves first. If you see a tree with plenty of brown leaves that is bare on one side only, that is likely to be the side the prevailing winds have come in from. The bare side of the tree is like a compass needle, pointing in the same direction that most gales come from.

Once you have noticed this broad effect a few times, try to spot how it is more pronounced with height. Winds are weakest nearest the ground and stronger with height, and this leaves a mark on trees in autumn, too. Look for totally bare high branches and almost full bottom ones on the sheltered side of the tree.

Now we're ready to look for the effects of local winds. The prevailing winds cover whole regions, but when they touch the ground, they change their behavior and give birth to lots of local winds.* If we notice that the leaves have fallen off the tree with a certain pattern, there will be a good reason for that. A combination of thinking about the prevailing wind direction, any recent gales, and the shape of the landscape around you may solve the mystery.

---

*I have written about these local winds in *The Secret World of Weather*.

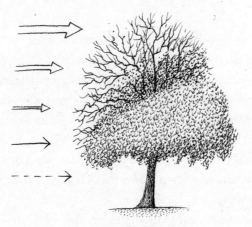
Autumn Leaf Loss Patterns

I was walking along a pavement in Fulham last autumn when I saw a line of three cherry trees. Two still had their leaves, but the one in the middle had lost them. It looked lonely and bare. After pausing to survey my surroundings, I discovered that the prevailing wind had been channeled down a street, then squeezed through a gap between two houses and plucked the leaves off the tree in the middle.

SHY FLOWERS AND SHOW-OFFS

A few trees flower before their leaves come out—the blackthorn is famous for its white blossom on bare black twigs—but most deciduous trees follow the progression I mentioned earlier: bare, bud, burst (leaves), flower, fruit, and fall. Once the leaves are out, we can keep our eyes sharpened for flowers. And our success in spotting them will depend more on ancient history than on the trees.

There are two main types of flowers on trees, and to understand what we're seeing and why, it helps to go way back in time. Conifers evolved before broadleaves and their reproduction depended on the wind carrying pollen from one flower to another. Later in the evolutionary story, some enterprising plants discovered that flying animals, mainly insects, could do a more precise job than the wind of carrying pollen between flowers. This led to huge differences in the flowers we see.

Imagine you're a tree that relies on the wind to carry pollen. It doesn't matter what your flowers look like—wind is inanimate: It doesn't make choices or show preferences. However, if your plan is to enlist the help of insects, like bees, you're suddenly in the business of attracting creatures that make choices. You're now in competition with all the other animal-pollinated flowers out there. If you don't make your flowers attractive enough for the bees to flock to them and take your pollen in preference to your competitors', you fail to reproduce and the show is over. Animal pollination is much more efficient than wind, but you have to make sure your flowers stand out. It's like the most ruthless horticultural competition imaginable: If you fail to get a gold medal, your family dies.

Most conifers are wind-pollinated and most broadleaves are animal-pollinated. This is why the two types of tree flower look so different. But we don't need to identify the tree or look at its leaves to work this out: We can just look at the flowers.

If you see flowers on a tree that are attractive, with pretty petals or eye-catching colors, you are looking at an animal-pollinated tree and you're in good company. Have a look around and you'll spot some insects taking a keen interest, too.

The same logic applies to scent: The wind has no interest in smells, but insects do. Any flowers that emit a scent are almost certainly pollinated by insects, because birds don't rely on a sense of smell. Plants with small flowers on wide flat umbels, like a waiter holding up a tray on the fingertips of one hand, aren't very pretty, but they're usually covered with small flies. They swap prettiness for a rich scent that mimics natural delights, like dung or decaying flesh, which is attractive to flies but can smell unpleasant to our noses. Hawthorn and elder trees have umbel flowers and one of the more agreeable scents. They often have dark specks of insects about them, too.

Wind-pollinated trees don't hide their flowers, but they don't stand out—there's no need for that. The wind carries pollen from conifers without any fuss, and most of the millions of pollen grains take flight all around us without anyone noticing. Once in a while you will spot conifers releasing pollen in thick yellow clouds known as the "sulfur shower," but that is the exception. Most of the time, few people spot the diminutive conifer flowers and only hay fever sufferers detect their pollen on the breeze.

Maples are interesting in-between examples. Their flowers are odd, with very interesting and ornate shapes but inconspicuous colors. They rely on both wind and insects, and their unusual flowers mark the transition from the old, wind-pollinated way to the new animal-pollinated era. Stop to look at maple flowers and you're staring at a bridge that spans millions of years of evolution. There will be a reason for any interesting shapes we see in flowers and I use a line to remind myself of this, "Where there's bells, there's bees."

Any flower with a pronounced bell shape is trying to stack the odds in favor of certain animals. This is very common in shorter plants: Foxgloves have long bell-shaped flowers that have evolved to favor bumblebees. They even have a pretty pattern on the lower lip of the bell that attracts the bee and acts as a landing zone. There is an intricate relationship between all flower forms and animal behavior. When scientists studied the flowers of one particular wild plant (*Erysimum mediohispanicum*) they found that the wider the petal on the flower, the smaller the bee it attracted and vice versa.

Some plants rely on birds as their preferred pollinator, which leads to distinctive shapes in their flowers. In their native South America, fuchsias attract a specific species of hummingbird, which uses its long beak to access the nectar at the end of tubular flowers. Flowers that attract birds are typically red in color, one popular theory being that birds can see red more clearly than bees, but the science is more nuanced and complex than that. Bird pollination is more common in smaller plants than trees, but the Indian coral tree lures birds with its bright red colors and generous helpings of nectar.

Every color has meaning, and the signals can change over the course of spring—many plants have flowers that don't stick to one color. The horse-chestnut tree has distinctive pyramids of flowers that send different signals through their short lives. The nectar-carrying parts of the flowers are white to start with, but turn yellow as they open, advertising their readiness for pollination. Once pollinated, the color changes to crimson, which is hard for bees to see. It's the flower's way of saying to the bees, "Move along, please, job done, and there's no nectar for you here anymore."

Last May I spent half an hour walking around the horse-chestnuts in the public gardens in the heart of Stratford-upon-Avon, in the West Midlands region of England. There was a definite trend in the colors, many more crimson pollinated flowers on one side of the tree than the other, but I couldn't decipher the reason. The investigations continue and, despite a failure to solve the puzzle, I still regard it as an excellent use of my time. If only every half hour in life were filled so well.

UGLINESS IS POPULAR
If we see flowers with beautiful petals on trees, animal pollinators are nearby, but they also point to a landscape clue.

Attracting insects depends on light and some openness—there is no point in any plant producing stunning large flowers in the heart of a dense dark forest. Wind pollination is unfussy: So long as a breeze can reach a tree, it will work, however dark it is. Flowers with petals are more common on trees that are isolated or grow in small groups, like fruit trees. Wind-pollinated flowers are more common in forests.

This means that we can use a rough rule of thumb: the bigger and prettier the flowers, the more open the land; the smaller and less conspicuous the flowers, the more likely we are in or near dense woodland. Pretty flowers are more popular with animals and humans, but wind-pollinated flowers dominate vast areas, especially above certain altitudes in each region.

FLOWER COMPASSES

Any part of a plant that has a relationship with light can be used to make a compass. Flowers with petals reflect light toward insects, which is why they are commoner on the brighter, southern side of trees. It is also why they are oriented toward the sun. Like the leaves we looked at earlier, many flowers aren't static but will twist to track the sun over the course of the day.*

*These days, many smartphones have a time-lapse option in the camera. If you set yours up to film a few daisies on the lawn for an hour or more, you will see this motion very clearly. The effects are most dramatic near the start or end of the day when the flowers will open or close as well as track the sun's movement. Or you can watch one I filmed here: naturalnavigator.com/news/2020/04/daisies-opening-a-time-lapse.

Where a clump of trees is growing close together, which is common with some species, like cherries, this effect is compounded. The southern side of the southern-most tree receives plentiful light, but the north side of that same tree receives hardly any: It is on the opposite side of the southern sun *and* it is shaded by its neighbors. One side of this tree will be covered with flowers, while the opposite side has barely any.

FLOWERS AS ARCHITECTS

It's time to revisit a sign we first met in the Missing Branches chapter, only this time, we'll bring the flowers to center stage. There is a small tree I pass daily, although to call it a tree is to pay it a compliment: It isn't much taller than me. It's not surprising that I pass it every day—it's very common along the paths in the chalk hills I live among. The clue is in its name, the wayfaring tree.

In spring, the wayfaring tree puts out wide umbels of scented white flowers, and in summer, red flattened berries appear that turn black as the season matures. When either the flowers or the fruits are out and the tree has its full leaves, it's pretty. It has a rounded shape, with some order and a little discipline in its form. But come winter, it looks like one of the most untidy shrubs in the land, with skinny branches heading all over the place. Chaos reigns. There is a reason for the mayhem, and it lies in the position of those white flowers from earlier in the year.

Flowers have a big impact on the shape of a tree. The two main jobs of a branch are to put out leaves to harvest energy, then to put out flowers and fruit to procreate. We have seen how leaf buds give us a clue to the shape of the tree (opposite buds mean opposite branches, and alternate buds mean alternate branches). A similar but slightly different clue is to be found in the flowers. It is always worth checking where they are positioned on a branch.

Each tree has to pick one of two strategies: It can either put the flower at the growing tip of each branch or it can grow flowers off buds along the length of the branch. It is tempting for the tree to position the flowers at the very tip because this will be the part of the branch that gets most of the light and is also most exposed to flying insects. So, why don't all trees do that? Because there is a problem with growing a flower at the very tip: A flower is the end of the road for that branch. A flower at the tip doesn't kill the branch, but it means it can no longer grow from that tip: It has to change its heading and fork off in a new direction. Each change in direction introduces a weakness to a branch and limits the total size of the tree before branches start to break.

Trees with flowers along the sides of the branches grow straighter; those with flowers at the tips of branches grow in a zigzag pattern. We can look for this effect from either end of the process: if it is spring and the flowers are out, we can see how trees with flowers at their tips, including magnolias, dogwoods, and maples, have a ragged, untidy appearance in winter. Or, as I like to remember it:

> Branches with flowers at their ends
> Are all forks, zigzags, and bends.

Flowers are reproductive organs and reproduction is a mature activity. The youngest trees don't reproduce so don't put out flowers. This is one more reason why younger trees look more orderly than older ones.

FRUITS AND SEEDS
All pollinated flowers aim to produce fruits and seeds, but the way that happens varies. Predictably, the biggest difference is between broadleaves and conifers. You'll be familiar with many of the fleshy fruits of broadleaf trees: We find them in supermarkets—apples, peaches, pears, apricots, and many more. And there are some you'll know well, but may not have thought of as fruits, like walnuts. In truth there is such an extraordinary variety in the fruits and seeds of trees that it is a little harder to look for broad patterns, but here are the few I've enjoyed noticing.

Cones are the fruits of conifers. A casual glance tells us we're looking at a cone, but it takes a while to appreciate the many individual shapes and peculiarities. Conifers have male cones that produce pollen and female cones that produce seeds. When we refer to a cone, though, we nearly always mean the female because male cones tend to be smaller, softer and less cone-like in color and shape. (In the appendix, I delve a little deeper into different cones.)

It's easy for us to see the logic of flowers reflecting light back to insects and growing more abundantly on the open and sunny sides of trees, especially the southern side. It's obvious, but easy to overlook, that fruits and seeds come from the flowers, so we will see more on the open, brighter sides of trees, too.

I love noticing how the red haws, fruits of the small hawthorn, cover the southern side of the trees in a long bright line at the southern edge of woodland. The fruits not only paint south with the broad brush of their color, but each one points close to south, too.

As always, there is an art within an art. Some fruits and seeds will appear more generously on the south side, but then the wind stamps its mark. Each February the male flowers of the hazel hang as catkins, and this is when I enjoy looking for the "hazel flag." There are strong winds in late winter and early spring and the catkins are tugged downwind, then hang on the opposite side of the branches to the direction the last storm blew in from.

MAST YEARS

The annual cycle of seasons is the one we're most familiar with, but longer and shorter cycles roll alongside them. Many trees that produce large seeds, including beech, oak, and hazelnut, do not produce the same number of seeds each year. Every few years there is a "mast year" when the tree releases vastly more than it did in the years either side. The timing of mast years is shaped by weather, but it happens because of animals.

Trees can survive without producing offspring every year, but animals have to eat regularly. The trees have learned through evolution how to use this to their advantage. If an oak tree dropped the same number of acorns every year, then foraging animals, like wild pigs, would fatten up on the acorns and breed successfully until there were enough pigs to eat every single acorn on the forest floor, in theory at least. However, if the oak tree plays a more cunning game and goes a couple of years without dropping many acorns, the pigs starve and the population reduces. (It does this in two ways: Some animals starve to death, but animals also have fewer offspring in lean times.) Then, boom, the following year, the oak drops a massive number of acorns and there aren't enough pigs to eat its copious bounty. More acorns survive and start life as seedlings. More simple genius from the trees.

JUNE DROP

Walk around an apple tree in early to midsummer and you may think it's in trouble. There is likely to be a good sprinkling of small fruits littering the ground under the canopy, if the animals haven't beaten you to them.

Apple trees drop quite a few of their early fruits, long before they have grown fully or ripened. This natural process is known as "June drop" and continues for several weeks, peaking a couple of months after flowering, typically in July. Many other trees, including citrus and plums, do something similar and it's no cause for alarm. I think of it as part of the "shotgun" approach to reproduction.

Plants produce vastly more flowers, fruits, and seeds than they would need to if every single one led to an offspring. Seeds can't start from fruits that never grew, and there can never be more fruits than there were flowers. A tree can self-prune any part it likes at any stage, so long as it has grown in the first place. But it can never turn back the calendar. It makes sense to grow more at each stage than will ever be needed, then to cut back. If all the tiny apples on a tree in May grew into full-size fruits, the tree would struggle to feed or support so many and has no need of them anyway. It's better for the tree to have fewer healthy, well-fed fruits than lots of starving ones. Hence the drop.

Muhammad Ali fought George Foreman in one of the most famous boxing matches of all time, labeled "the Rumble in the Jungle" by salivating promoters; it has been claimed by some as the "greatest sporting event of the twentieth century." The match took place in the enormous capital city, Kinshasa, now in the Democratic Republic of Congo but then in Zaïre. (Kinshasa is a city and definitely not a jungle, but it was nearer to such than Las Vegas, which was good enough for the marketing people.)

Ali was the clear underdog and went on to win the fight using a risky novel tactic nicknamed "rope-a-dope." Ali retreated to the ropes, letting Foreman think he was well on top of the match, then Ali held up his guard as Foreman unloaded a barrage of punches. This exhausted Foreman, who was surprised to watch Ali charge back into the match late on and beat him.

Each spring, as the soft new leaves emerge from the buds, the caterpillars and other animals launch themselves at them, like George Foreman. The trees can lose almost all their leaves during this onslaught. But they take the punishment, wait on the ropes and then, later in the season, come back strongly with a wave of fresh new shoots and leaves, called "lammas growth." (Lammas is a Christian festival celebrating the first fruits of the harvest and traditionally falls on August 1 in the northern hemisphere.)

Oaks, beeches, pines, elms, alders, firs, and many other species put on a second spurt of shoot and leaf growth well after spring, nearer midsummer. They come off the ropes fighting. Interestingly, these late lammas leaves can have a different form from the original spring ones. Oak lammas leaves are skinnier with shallower lobes.

THE TEN STAGES OF LIFE

Each time we look at a tree we gain a sense of how old it is. We know its size gives an instant clue and measuring the circumference a slower methodical one, but there are lots of other signs, too. We see them, but don't always register them.

In 1995 Pierre Raimbault, a French arboricultural scholar, decided that a tree has ten stages of life and that we could read these by noticing certain morphological traits. At the earliest, Stage One, the tree is obviously very small, but the key thing is that there are no side branches at all. By Stage Two, the tree has branches, and by Stage Three, there are secondary branches off those.

At Stage Four, the tree has pruned the lower branches that are shaded and ineffective. Between Stages Five and Six, the tree prunes more aggressively and the lateral branches grow more determinedly. This leads to a change in appearance, a broader canopy with a clear gap below, where the shaded lower branches once were. By Stage Seven, the trunk is bare below the canopy.

Trees get taller for the first seven stages, but then the canopy starts to collapse and the tree loses height, even as the trunk continues to fatten. By Stage Eight, the tree has stopped growing from its extremities and now rejuvenates in parts closer to the trunk, leading to a halt in canopy growth before a slow retreat begins in Stage Nine.

You'll recall that the tree has epicormic buds under the bark of the trunk, which wait patiently in the shade for their turn. The stress of aging means the leaves no longer grow at the extremities and now light reaches the trunk, which triggers growth in these buds. It may have taken centuries, but they get their day in the sun, literally. (The stress of aging also changes the hormones in the tree, leading to the new growth.)

The trees that make it to Stage Ten, and many don't, start to collapse in on themselves. The tree is still alive at this point, but crumples and relies on those new lower shoots from the trunk for survival.

Of course, these are just Raimbault's divisions. We could add some of our own or ignore a few. And trees, like people, age differently if they've lived a tough life—the gnarly dwarf trees on the thin, exposed soil of Wistman's Wood in Dartmoor, in South West England, don't readily reveal their real age. But it is still an interesting exercise to look at a tree and try to pick the stage it's reached, as Raimbault would. It's like trying to read a distant town clock through mist: Sometimes the number is obvious; at others it's harder to make out.

And the clock doesn't stop when the tree dies. The trunk of a mighty beech near our home snapped about 33 feet (10 m) off the ground and came crashing down. It must have harbored a weakness for many years. Most likely a hole in the bark let in a fungus that had slowly devoured the strength of the stem. It came down about five years ago and, rather wonderfully, the larger part on the ground managed a whole spring and summer of growth in the year after it fell. It had no connection to the roots, but there was enough energy in the trunk, branches, and buds for the leaves to grow for one more season.

TREE CALENDARS AND WOODLAND CLOCKS

Conifers grow a layer of branches each year. This means we can gauge the age of a conifer by looking for these levels or "whorls" of branches. It's easiest in younger trees, up to about ten years old, as the gaps between each annual layer are more obvious—young fir trees show this effect strongly. Things get more crowded and harder to spot as the trees mature, but the same principle still applies.

Young branches grow from their tip, led by the terminal bud. The process is similar to the way the top of the tree grows, only outward. The growth is seasonal and its stop-start nature leads to scars forming around the twig. The length of the gap between each scar marks the growth for that year. It varies with the age of the tree, the conditions during that season, and earlier ones: A young branch on a young tree in perfect conditions (a good balance of ideal temperatures, light, and rain) will lead to longer spaces between the scars.

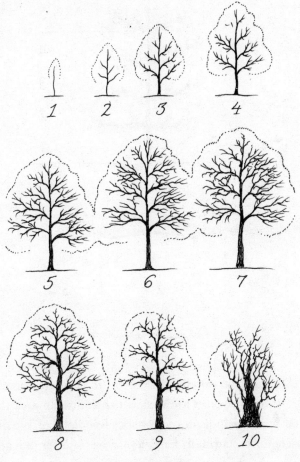

Raimbault's Ten Stages

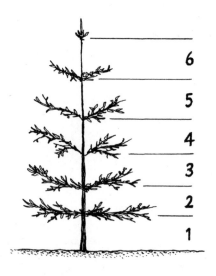

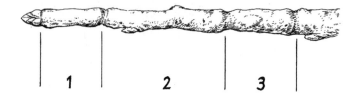

Conifer whorls and bud scars reveal the annual growth in trees.

The age of woodlands can be read in the smaller plants. Some species are so slow to colonize an area that their presence indicates continuous woodland on the spot for centuries, normally for as long as records for that area have existed. These niche plants are known as ancient woodland indicators (AWIs). I'm fortunate to live near many ancient woods and see wood spurge, butcher's broom, and many other ancient clock hands regularly.

Time changes woodland. The early decades are marked by an initial light-grab phase, where every tree is out for itself and tries to take as much of the available sunlight as possible. But this leads to an unsustainable number of trees and the problem only grows as they mature, their canopies rising and filling out—many will lose the struggle and die. A mature woodland has fewer trees and fewer species of tree than a young one. This is one reason why managed woodlands are thinned period-ically. As I write this, I can hear the dull, distant rumble of heavy machinery carrying timber out of the woods. People instinctively see this as abhorrent, vandalism of our precious woodlands, but it is good for them, leading to healthier remaining trees and greater biodiversity.

There are interesting cycles in the organisms that live alongside the trees. Much of the most interesting animal life thrives at either end of a wood's life. In the early days, there is plenty of light, which brings more small plants, insects, and birds. As woodlands grow old, a lot of decay-ing dead wood will lie on the ground: It leads to a spike in insect life, helping so many other species.

If you are on a winter walk through broadleaf woods, you'll notice that you can see so much farther. It is a little ironic: We think of summer as the season of light, but in broadleaf woods it can be oppressively dark under a full canopy. In winter, the sky may be a leaden grey, but the light pours down among the trees. Mosses, ferns, lichens, and liverworts love the moisture of winter and make the most of this light, continuing to grow if temperatures allow. After a mild spell, you may notice a greening of the woodland floor and the lower parts of the trunks or branches, like a mini spring in the dead of winter.

12

Lost Maps and Tree Secrets

*I'll Meet You by the Tree • A Desire Map • A Valley Is a Gift •
The Tree Pair Secret • Trees and Clocks • Emperors of Time •
Birds and Treesong • The Miniature Map • Two Journeys*

I'LL MEET YOU BY THE TREE

I have an old habit, which is ancient in humans and older still in animals. I like to make sense of new landscapes by identifying landmarks. A landmark is any visible object—natural or artificial—that helps us understand where we are. Trees have a long history as local landmarks. Any proud isolated specimens force their way into local lore because they tick the boxes that all good landmarks must: They are unique, recognizable, long-lived, and they stand out.

> Abram traveled through the land as far as the site of the great tree of Moreh at Shechem. At that time the Canaanites were in the land."
> —Genesis 12:6

Who has never arranged a meeting by a particular tree? Landmark trees don't need to be isolated, but it helps if they're conspicuous. There is a pair of impressive beech trees in my local woods: They're about a hundred years older than the neighboring trees and stand proud, unmistakably grander. In our family this pair earned the nickname of the "Elf Trees" because when our sons were younger they could often find coins tucked among the gnarled roots at the base of the trunks. My wife and I explained, straight-faced for longer than the Santa stories held, that the elves must have left them. The elves had a strict code: They only ever left coins, never paper money. The game continued for about five years and only stopped when the kids started staring at our hands longer than the tree roots.

Most people notice obvious or dramatic things in a landscape and often miss the subtler things. In the age of smartphones, some miss everything, but that's a different story. Take a few seconds to think of the following cities in turn and try to picture them in your mind's eye: Paris, London, San Francisco, Agra, New York.

There's a good chance that at least one of the following images entered your mind: the Eiffel Tower, Big Ben, the Golden Gate Bridge, the Taj Mahal, Times Square. . . . The less well we know a city, the bolder the landmarks we need and vice versa. A person who has lived in the same city all their life will use much cuter landmarks: "Meet me by the pink graffiti."

When people are new to a city, they tend to refer to something very bold, and by very bold, I mean clichéd. In extreme cases the landmark becomes more famous than the city itself.

The same rules apply with nature. Everybody spots the mighty solitary oak, but few notice the thorny sapling they passed earlier. If you walked in our woods, you would notice the elf trees—they stand out. But I've spent so many days in those woods that I could happily list dozens of tree landmarks. I've got names for many, including long-dead trees. There is the growth (epicormic) spurting up at either side of a stump—the Viking Helmet; then there is the inverted beech stump with weathered root bases pointing upward— the Crown; or the decaying branches that are arched over to the southern sky—the Claw. I'm confident that these remain invisible to most passersby, not least because I've watched it happen.*

You will have your own landmarks in the places you know best. But the question is: How can we start to notice these features when we're just visiting? There's a simple technique that works: Imagine you plan to meet someone near the spot where you're standing in a couple of hours' time and you have to describe that location only by reference to the trees. Better still, meet someone there: It will add an edge to the exercise and place those trees firmly in both of your memories.

---

*Photos of these landmarks can be viewed at naturalnavigator.com/ news/2021/03/what-is-a-landmark.

This is a fun exercise in a city park, but the same task a long way from a path in a large wood starts to feel more "edgy" and certain previously invisible features shine out.

A DESIRE MAP

Expect to see trees. If somebody kidnapped you, blindfolded and then carted you off to some random patch of land, the first things you can expect to see on removing the blindfold are trees. Left to its own devices, nature comes up with trees in most situations because they win the harsh game of survival in all but the most extreme environments. Nature's motto: "Unless otherwise instructed . . . trees."

If we turn this logic on its head, it gives us a clue. If we look at a landscape and see no trees, we can say with confidence that there is something unusual about human desire for that land. There are no trees on the tops of large mountains, in the oceans or in hot or cold deserts. Humans have little desire to live there and trees can't survive there either. But in almost all other landscapes, we should expect to see trees growing.

The land that is most desirable to humans is home to few trees: They have been displaced by cities, the concrete jungle. But even in rural areas we have been stripping the land of trees to make space for homes and agriculture for the past ten thousand years or more. Trees that survive in farmland have escaped the plow. Farmers left unproductive land, anything that was too steep, rocky, or otherwise unworkable, to the trees. Near my home there is a series of steep gullies, probably carved by the torrents of glacial meltwater from the last ice age. These steep-sided valleys have always been too much for horse-drawn plows or even the impressively agile modern tractors. They are lined with ancient woods.

As soon as I head south and uphill from my home, I am plunged into beech woodland. Some ashes and maples are sprinkled between the beeches, especially near the edges, but there is no doubting that I am walking through a landscape dominated by broadleaf trees. A couple of hours later, having descended off the chalk ridge of the South Downs, in South East England, I expect to see change because I know that the rocks and soils are different on the north side of the hills. And as soon as I see the broadleaves disappearing, replaced by conifers, I know I've reached Graffham Common. The soil is noticeably sandy underfoot, too dry and harsh for agriculture. It was left to scrub, conifers, and commoners.* Or, as the writer John Lewis-Stempel puts it, "Conifers mark poverty."

---

*Graffham Common. The name is a clue: Common land in the UK refers to the historical right of the whole community, the commoners, to graze cattle on the land. It wasn't—and still isn't—owned by a private individual or company but is shared by all. It's not common land thanks to the philanthropy of earlier generations, but because the soil is sandy, making it too nutrient-poor for agriculture. Each generation is generous with the resources it doesn't need.

If we don't see trees, it's because lots of people want the land or nobody does.

A VALLEY IS A GIFT

Every time you step into a valley, there are certain guarantees. There will be higher parts and lower parts; there will be at least two slopes with different aspects and they will receive differing amounts of sunshine, rain and wind. Nutrients are washed downhill, making for richer soil lower down. And the trees will reflect all of this. No trees can cope with all environments: They have to specialize. Each species has evolved to thrive in certain habitats. They have their niche, which means they are all fussy about certain things, especially light, water, wind, temperature, nutrients, acidity, and disturbance. Many trees are comfortable near the middle of some of these variables, but each will be particularly sensitive about one or more. That is what gives it an edge in its preferred environment. Whenever you find yourself with a view of a valley, accept this as a tree-reading gift and look for changes.

THE TREE PAIR SECRET

Kingley Vale is a nature reserve near my home that is nationally renowned for its grove of ancient yews. Some of the specimens are mighty, weird, and slightly spooky. On a cool, sunny October morning, I visited Kingley Vale and placed my hand on the memorial stone near the top of the reserve. It was a small salute to an interesting man. The words on the metal plaque read:

> In the midst of his nature reserve which he brought into being, this stone calls to memory Sir Arthur George Tansley, FRS, who, during a long lifetime, strove with success to widen the knowledge, to deepen the love and to safeguard the heritage of nature in the British Isles.

Tansley has been lauded for conservation work ahead of his time, but he is a hero in another important way, too, one that is very relevant to our understanding of trees.

Early in the twentieth century, Tansley took inspiration from the work of the Danish botanist Eugen Warming and pushed a particular branch of ecology forward on the international stage. In 1911 he helped organize the first International Phytogeographic Excursion. The "international" and "excursion" parts are straightforward enough: This was a gathering of scientists from Europe and the USA who ventured out into the British Isles for research. But the middle word, "phytogeographic," is the key for us. "Phyto-" as a prefix means any study that relates to plants. Geography is a much more familiar word, but not the easiest to define.

About a decade ago, I was fortunate enough to have a conversation with the then director of the Royal

Geographical Society, Dr. Rita Gardner. I made a confession: "This is a bit embarrassing. I'm a fellow of the RGS, but I couldn't define geography if I tried. What is geography?"

My question was sincere and hopefully polite, but it was fueled by a slight unease at how geography had morphed into such a vast and sprawling discipline during my lifetime. My personal, imperfect, and old-fashioned view was that a subject rooted in the study of physical processes, like glaciers and volcanoes, had come to include urban planning and income distribution charts. But Rita Gardner didn't earn seventeen initials after her name (CBE, FRGS, FRSGS, FAcSS) without knowing how to deal with problems like me.

"At its core, geography is the local study of change."

So, phytogeography is the study of how plants change with location. Or how they make a map. In our quest to find meaning in trees, Tansley played an important role, pushing forward a fascinating branch of plant science. It is a branch that hides in front of our eyes and reveals how we can all make a map using trees.

I think everyone senses, even if unconsciously, how we regularly see certain plants near others. Many adults recall and pass on the childhood lesson: Rub a dock leaf on a nettle sting. And, rather obligingly, there always seem to be dock leaves growing nearby. It is tempting to see kindness in this arrangement, but the truth is simpler: The two plants favor the same habitat niche of nutrient-rich disturbed soil.

Every plant is trying to tell us something about the land around us. Whenever we see a single plant, it gives us a probability of seeing others, but if we see two plants doing well in an area, it changes the probability picture dramatically. If I see nettles in my local area, there is a decent chance of spotting rough meadow grass, too, as they favor similar soil. But if I see nettles and dock, I am almost guaranteed to find rough meadow grass nearby—the probability shoots up massively.

What on earth has that got to do with reading trees? A lot. Bear with me.

Near the start of the book, we saw how a single tree species can be used to make a map. It's a very interesting map, but it's broad-brush. Once we're comfortable with spotting rivers using willows and poor soil using conifers, for example, we are ready to learn to read a much more detailed map, and we do that by noticing tree pairings.

Everywhere in the world that trees grow, it is possible to gain an extraordinarily detailed picture of our surroundings by recognizing the pairing of trees with one other plant. And almost nobody knows this little trick, which is why I call it the "Tree Pair Secret." There are so many combinations across the world that there's no point in any attempt to list them here. And besides, the aim is not to learn the words but to notice the popular pairs and patterns in your part of the world. Let me show you how this works by giving you some of my favorite examples from my walks in Sussex.

You will have spotted that I have referred to the beechwoods I live near. When I'm moving through these woods, I try to stay tuned to which plant is thriving among these trees: Which plant is the beech tree paired with? It is usually either dog's mercury or brambles, and they each tell a different story.

If I see a carpet of dog's mercury on the ground, it means I'm crossing an area where the beech has become dominant and thrown its shade over the ground, snuffing out most other trees and lower plants. I may spot the odd yew or one or two other shade-tolerating plants, but the variety will be severely limited by the deep shade. Near the edges of this woodland, I will find ivy, and in places where people have disturbed the woods there may be a sprinkling of birch or ash trees, but little diversity in the plant life. The heart of these woods can be a little oppressive in summer, even for the birds and insects, and they are quiet.

The second I spot brambles growing among the beeches I know I'm looking at a new pairing and the small world I'm walking through has changed. The land is brighter—brambles can't survive in deep shade—so I'm either in a place where the beeches are too young to blot out the sun or near a track or clearing that allows the light in. Nearby I will find holly, ivy, bracken, more mosses and possibly an oak, maple, or ash, too. The variety of plant life has shot up and the animals follow. The birds and insects are much busier among this pairing and it is noisier, almost bustling, at dawn and dusk on the bright side of the equinoxes.

Whichever tree dominates the land you walk through, look for its regular pairings: It will add many colors, sights, and sounds to your map.

TREES AND CLOCKS

One September afternoon I planned to walk in the hills, then meet friends in a village called Halnaker. There is an art to arriving on foot at the right time at a place a few hours away. Over the years I've learned that the logical approach doesn't work for me. Most sensible folk could be forgiven for following a simple formula: Look at the distance to be walked, estimate your walking speed, divide the first by the second, subtract that from the meeting time and . . . set off. But there is a minor flaw in this plan, and it's a major thorn for me.

This formula assumes certain things, and the most alarming is that there will be no detours, physical or philosophical. There is a word that describes a walk planned without any slack for pleasant distractions: appalling. As a reader of this book, you would relish, I know, the opportunity of killing time among the trees. Let us spare a moment to think of the poor souls who wouldn't understand what we're talking about. This is the Law of Time in the Woods: "An hour can be hidden in the trees, but never found."

It always makes sense to add time to the plan. The worst that happens is that we find ourselves nearing our destination with an hour to spare, and what a wicked-less pleasure that is.

I found myself in this luxurious situation on my way to Halnaker. I was descending a conifer-covered hillside when the sun told my watch that I was running early. Forced by this happy circumstance, I left the path and wandered past a private-estate sign that told me I shouldn't proceed. I forget the exact words, but the tone implied I'd be lucky to make it a hundred yards without being shot. I felt lucky.

I wandered up and down the slopes, my feet sinking into beds of soft dead needles, then negotiating harder bumps and stumps. Among the conifers, near the end of the day, the sun was sometimes visible between the foliage and the ground. I decided to toy with the sunset.

Sunset appears to happen later when we look downhill and earlier when we look uphill. (When we look downhill, it has the same effect as lowering the horizon: The sun has farther to go to reach it, so sunset is delayed.) By ranging over a series of undulating mini-crests, I played the sunset tape forward and backward. And by choosing to spend time among the pines instead of the spruces, the game was a lot easier: The sunset window is much bigger under pines that have shed their lower branches.

If you try this time game with broadleaf trees, you'll spot an interesting pattern. In any area with deer or other large mammals that feed off the leaves of living trees, look for the "browse line." Browsing deer are like fussy gardeners and create a neat line that marks the bottom of the tree's canopy. (An untouched canopy is uneven and undulates.) The browse line mirrors the contours of the ground because the animals reach the same height above it—trees on a slope have a matching slope in their foliage: the larger the population of animals per area, the scarcer the food and the neater the line.

The Browse Line

The browse line shines out on isolated trees, they look pruned against the sky, but we may overlook it in woodland. Deer and other grazers will clear the lower level of woodland, which improves visibility. In my local wood, large areas have been fenced off to protect saplings from animals. When I cross these fenced-off areas, the leaves reach much lower down and smaller plants rise up—it's like the trees and the undergrowth are squeezing the gap. Put another way, if you can walk freely and see a long way in broadleaf woodland, you're not alone. All of this means that the number of browsing animals in your woods changes the time of sunrise and sunset.

Before I knew it, I was back on schedule but had lost any desire to reach the village. Unfortunately, the spell that allows us to move the sun up and down in the sky does not magic our friends into the woods. I set off downhill, filled with thoughts of how I would play with the moon under the trees on the way home.

EMPERORS OF TIME

The age of trees has a dramatic effect on everything else we see.

I was once fortunate enough to spend an afternoon exploring the Knepp Estate in West Sussex with Isabella Tree, the award-winning author and joint creator, with her husband, Sir Charles Burrell, of the first large-scale rewilding project in lowland England. Their land sits on sticky clay and has proved a relentlessly horrible terrain for agriculture. Encouraged by their lack of options, Charlie and Isabella decided to let nature start making some of the decisions. It didn't happen like that, I don't think, but I like to imagine them howling into the wind, "Plants, you're not being reasonable! If you won't make the slightest effort to cooperate, we refuse to pick up the pieces anymore. You're on your own!"

It was a brave decision. Maybe a little like teaching teenagers that their bedrooms don't tidy themselves automatically by refusing to acknowledge the mess. (In our case, this tactic lasted only until the half-empty cereal bowls started their own rewilding project.)

Where once plowed fields refused to yield much joy, there is now a vigorous ecosystem that is regenerating itself. It is a landscape that tests each visitor's sensibilities. If you are the sort of person who likes perfect stripes in lawns and gets a bit twitchy when autumnal leaves make a mess, then look away: You're not ready for this. If, however, you like your nature unmarshaled, dive in: You'll love it.

The land is mostly open, broken in places by thick clumps of thorny scrub, willows, and some ancient oaks. Each of these plants marks a moment in tree time. There is an old saying, "The thorn is the mother of the oak." Among the brambles and other thorns there was a young oak, its bark only recently hardened. Thorny scrub protects young oaks from animals in their vulnerable early years.

And oaks were once so vital to the national interest that the scrub was protected, as Isabella had explained in her landmark 2018 book, *Wilding*. A 1768 statute decreed that anyone extracting thorny scrub would face three months in prison and lashes of the whip.

The thorns had done their job: We could see nibbled shoots at the edge of the clump of brambles, but the young oak at their center had survived long enough to grow above the danger. The brambles would get no thanks for this: In a few decades that same oak would steal their light and starve them of energy. Nature does not brim with gratitude.

Less than a minute's walk from the young tree, a mighty veteran oak was nearer the end of its journey. We stood in the shade of this ancient tree and I admired its grand eccentric form. One major branch had collapsed completely and epicormic shoots on the trunk had matured into good-size limbs—successful Plan B branches.

The organisms that grow on trees care about the species of their host, but many are also fussy about size and age: "There are some fungi that will specialize in a particular width of branch and will only grow on that size," Isabella explained. She pointed to a bracket fungus on the ancient oak and explained that it was the very rare, *Phellinus robustus*.*

---

*Isabella had learned about this fungus from a man called Ted Green, a legend in the world of ancient trees. A few years earlier I had been lucky enough to play second fiddle to Ted as we led a walk around the ancient oaks of Windsor Great Park, west of London, for a Woodland

It can survive only by passing from one veteran oak to another; it will not last in a landscape of younger oaks—a single example that demonstrates the fallacy that developers do no harm when they replant the same number of trees as they fell.

The goat willows attract an elusive and much-loved creature, the Purple Emperor butterfly, whose caterpillars feed and thrive on these trees. The mature insects are busy around the willows, too: "They spend a lot of time chasing females through them, as she'll be looking for just the perfect leaf to lay her eggs on. To secure their territory, they're so vicious. They really are extraordinary insects—they'll even chase birds."

"What?" I said. My face must have been a picture as my mind wrestled with the image of a butterfly mobbing a bird.

It's a butterfly with bizarre tastes, feeding on sap some of the time, but topping this up with dung and carrion, too. There are stories about aficionados of the Purple Emperor who are so desperate to see it close up that they use their own weird bait recipes to lure the butterfly down from the canopies to the ground. We had fun sharing examples that we had heard of: peanut butter, shrimp paste, babies' diapers, rotting fish, Brie, dog turds. . . . We stopped before we felt queasy.

Trust charity event. Like fungi and trees, nature zealots live in an interdependent ecosystem: We regularly find ourselves growing and learning in the same habitats.

The Purple Emperors mark their territory with aerial displays, and their favored places for this were around the top and downwind of the ancient oak tree: "They'll chase each other around the crown. It's like a lekking site for fallow deer."

Isabella told me that the butterflies need the willows, but the willows need patches of bare earth for their seeds to germinate. When the seeds fall from the trees in late April, it's no good if they land on grass or scrub: They need exposed damp soil to start life. And that is one of the reasons why there are pigs at Knepp. They rootle in turf, opening it up for the seeds of the willows, just as ancient boar would have done in centuries past. Trees are the emperors of time. In one delightful afternoon I had witnessed how young thorns, young oaks, maturing willows, and ancient oaks had shaped the landscape and played their part in the flap of the wings of the most ephemeral of butterflies.

BIRDS AND TREESONG

Our brains are permanently a bit flustered, but when we slow down to take a deep breath and a proper look, we find such extraordinary variety and richness. It's actually quite impressive how much we've managed to ignore.

One late spring afternoon, after a couple of hours' walking across the South Downs in Sussex, I sat on a chalky bank and had a long drink of water. Then I scoured the land all around me. It always amazes me how we see differently and often better when sitting down. It's illogical as we can see less far. On a flat landscape, we can see about 50 percent farther when standing up than we can when sitting on the ground. And if we see 50 percent farther, that means we actually see more than double the land area in all directions.

When sitting comfortably, we start to see things we didn't before. This is less to do with the physics and more about the psychology. My theory is that the multitasking workload on our brain reduces when we sit down—we're not sending so many signals to muscles and we're fielding fewer signals back from them that say nagging things, like "I'm tired. Put your weight on the other foot already." And maybe this frees some headspace for noticing things in a landscape. Try it with the next tree you pass. Look at it while you're walking, then look again after you've sat down. I promise you'll see things you didn't before. (Academics, please do write to me with a complex name for this simple phenomenon, perhaps kinetic tunnel-vision syndrome?)

From my rest spot, I was enjoying looking for birds among the thorns and spines. I noticed a robin hop up several levels, then start to sing from the top of a young maple tree. I might not have thought any more of it, but it made me curious: why do birds sing from the tops of trees? It makes sense when you need to use your eyes, but song is not about that. Why not just sing from the comfort and cover of a bush? Birdsong, like all sounds, travels farther the higher you sing. That's why church bells are at the top of the tower.

It must take some effort to reach high points to sing and to put heavy bells in the top of a tower, so it must be worth it.

The bird sounds we hear are connected to the height of trees and the pattern the trees form in a landscape. Birds are territorial and the territories they favor are usually mixed; most species prefer to avoid wide-open spaces and deep, dense woodland. Their perfect triangle includes some trees, some food opportunities in open ground and a water source.

By combining these two simple ideas, we find that birdsong and trees form part of the same map. If you're walking across open land and notice a clump of trees, the probability of hearing and seeing birds increases. The same is true from the opposite perspective: The sound of birds in open country indicates the likelihood of finding trees.

If you're crossing dense woodland, it's common to experience long periods without hearing birds, then notice a sudden increase in the sounds they make. That means you're probably getting nearer to the edge of the forest.

On any walk across varied terrain, the height and density of trees play with the sounds we hear and deserve to share the credit. Pause, close your eyes, and savor the treesong.

THE MINIATURE MAP

Trees make a large map by reflecting the environment they inhabit. But they are not passive: they leave their own marks on the land. By learning the habits of each tree, we can predict certain changes very close to them.

Some of this is common sense, but much isn't. And very few take the time to notice any of it.

Each tree has its own "shade profile." The shape, depth, and timing of the shade each tree casts is unique. Spruces cast a deep shade over a narrow area; oaks cast a moderate shade over a broad area; aspens let quite a lot of light through. Ashes come into leaf late; elders are early. There are good reasons for these differences, which we explored earlier. Now we will focus on how trees' shade habits dictate the other plants we find close by. In summer, I often find wildflowers under birches, but next to none under yews. And I find early flowering woodland plants, like bluebells, under late-into-leaf trees, like beech, but few under the early-leafing elder.

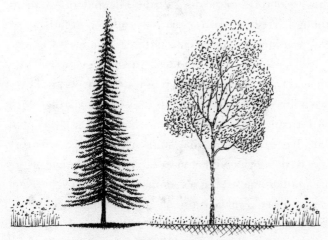

The shade cast by each tree changes the world beneath it.

Trees cool the air and the land beneath them, through shade, of course, but also wind and evaporation. A breeze accelerates under any isolated tree because the tree causes pressure differences in the airflow. A tree loses water through its leaves—transpiration—which also cools the air beneath it. Through shade, wind and transpiration, each tree has its own "cooling profile." A study in California found that urban trees can reduce the need for air-conditioning by 30 percent.

Each tree sheds its leaves in a unique way, and each leaf has its own life once it falls to the ground. Some decay quickly, others linger; some are especially rich in nutrients, others less so. Beech leaves are rich in nutrients, but they absorb most of them as few other plants can make any use of them in the deep shade. The spiders love them, though. Trees in towns can't recycle the nutrients, so we find leaf litter on pavements and trees that need feeding.

Alders change the land near them in a couple of very special ways. They are a sign of water nearby, and if you see a line of them, you are probably looking at the course of a stream. In this sense they are mapping the larger area, but they also change that landscape. As we saw earlier, alder roots help protect riverbanks from erosion, acting as a buffer against the soil-munching habits of the water. This can help make sense of some of the patterns you'll see where alders line the banks. All natural rivers and streams curve. If alders establish themselves anywhere on the banks, they will interrupt this natural flow, leading to more intricate wriggling patterns.

The second way in which alders change the land is thanks to their rare ability to "fix" nitrogen from the atmosphere. All plants need nitrogen compounds, but most rely on their roots finding enough in the soil. Alders have formed a partnership with bacteria that can take up nitrogen directly from the air, where it is always plentiful. Have a look at the roots of alders—their watery habitat can make this easier—and you should be able to spot nodules on the roots where the bacteria work their magic. (I've enjoyed looking for them for decades, but only ever noticed their shape. I'm always shocked and delighted by the bold simplicity of what I can fail to see. It was only after a recent conversation with an ecology expert that I learned to pay more attention. Dr. Sarah Taylor, a lecturer at Keele University, revealed, "Alders have a symbiotic association with Frankia bacteria—they look like mini-cauliflowers have attached to the root system. A red color means the bacteria are active. A dull grey-brown means it has died off. In eroded riverbanks I love looking for these structures—they're otherworldly.")

In this partnership the tree gains the nitrogen it needs and in turn feeds the bacteria with sugars. This is great for the individual tree, but it's also very positive for the land. Alders can grow in areas that are too nitrogen-poor for other trees. Then, when they shed their nitrogen-rich leaves, this fertilizes the soil and makes it viable for other species to follow. They are pathfinding trees, paving the way for others. Or, as the poet William Browne put it in 1613:

> The alder, whose fat shadow nourisheth,
> Each plant set neere to him long flourisheth.

The lights on a Toyota Hilux flashed and I flashed back. The driver had recognized my car from its simple description: a black Land Rover. I pulled in, did a three-point turn and followed the Toyota out. I tailed it deep into a wood in the Wiltshire countryside, in South West England, and we only stepped out to introduce ourselves where the country lane stopped at a locked forestry gate.

"Colin, I hope?" I reached out my hand and was relieved to receive a warm greeting. If I had misread the flash from the car lights and followed a stranger so deep into the countryside, there might have been an awkward conversation.

On a cool March afternoon, I had driven into the village of Sixpenny Handley in Wiltshire, where I had arranged to meet Colin Elford, a lifelong woodland ranger and author. We had been introduced by another ranger we both knew when we discovered that we liked each other's work.*

---

*I had very much enjoyed Colin's book, *A Year in the Woods*, and he told me he had really enjoyed my *Nature Instinct*.

We walked into Cranborne Chase, an Area of Outstanding Natural Beauty, and Colin began to explain how he managed the land for the benefit of all species, not just one or two. But he put it more eloquently, in a rich Dorset accent, "I don't like to think of just one species because then you bugger it up for all the other things you don't know about."

Colin explained how he cuts and "drifts" the hazels, dropping and laying the trees down in a way that leads to a mixed habitat on the slope below. This creates ideal habitats for a number of different species, including summer nesting areas for dormice. We passed many plants that I know from my home patch, wildflowers like violets and herb Robert, as well as one I rarely see, the parasitic Toothwort, a pinky-cream spike of not-pretty-not-ugly flowers that feed on the roots of the hazel. The undergrowth was thick with the sound of squirrels rippling away from us. In the mud, I saw the tracks of Fallow, Roe, and Muntjac deer.

We looked at tree branches that had come down in recent gales and traced their stories back through fungal decay at the break to squirrel damage that had let in the fungi years ago.

There is a twin joy in these days for me: I get to share time with a kindred spirit in a beautiful place and I always learn something golden. It is a treasure hunt, but I rarely know beforehand what the treasure is. I knew that someone of Colin's experience could teach me lots in his backyard, and he did. All afternoon I reveled in reading the landscape through the eyes of the person who knew it best in the world. It was early evening and we had been exploring for several hours when Colin led me to the gold.

We passed Neolithic burial sites and Grim's Ditch, earthworks dug about 2,300 years ago that once formed the boundary marker between Iron Age tribes. The area is rich with features that archaeologists savor, but the treasure lay deeper in the woods, just off a muddy track. We made our way between the trees and, as we stepped between more hazels, Colin told me a story about a local gamekeeper who was locked in a fiery tussle with poachers. Friction grew, and the gamekeeper ended up losing the fight and his life. Poachers, like pirates, have a colorful reputation, but black is never far from the palette. (I was recently crossing a woodland in the low light of late dusk and my shin caught on something hard in the undergrowth. After switching on my headlamp, I saw a metal crossbow bolt stuck hard into an old log—a sharp sign of deer poachers. If I had timed the walk differently, the pain in my leg might have been a lot worse.)

The woodland floor was thick green: Dog's mercury, nettles, ramsons, and a dozen other species spread away from us as far as we could see. And then the land changed suddenly and dramatically. Beneath a pair of walnut trees, the small plants gave up and the soil was totally bare. As far as the eye could see, in all directions, the forest floor was carpeted with the leaves of small plants, but now we stood next to a dark island of mud without a splash of green or leaves of any kind. The bare earth stretched as wide as the bare branches of the walnut trees. I paced about, taking photos and talking excitedly. I knew what we were looking at, but I'd never seen such a perfect, beautiful, striking example.

Trees are not cuddly green saints. At the genetic level, all nature has a selfish agenda and some tree species show this more ruthlessly than others. There is a botanical phenomenon called allelopathy, whereby plants produce chemicals that poison or inhibit other plants nearby. Many sharp-elbowed shrubs and trees display this habit, including the rhododendron, horse-chestnut and black walnut. They have a reputation as toxic neighbors.

Allelopathic trees aren't careless psychopaths looking to kill everything within range. They are more targeted, secreting chemicals that are most effective against certain other species. The black walnut infuses the soil around it with the poison "juglone," which is especially lethal to the trees it is most likely to be competing with, like birches. If you see a tree with bare soil around it, but one that doesn't cast a dense enough shade to explain it, you may be looking at a patch of poisoned earth.

Trees give us a clue to the types of animals we will see nearby and many make their homes in nests or among the larger roots. Elder trees are common at the edge of villages and animals such as rabbits like to make their homes near them. I remember it with this: "The elders of the village know where the animals live."

The forest floor looks and *feels* different from open land because trees always change the upper levels of the soil. Each woodland feels unique thanks to the trees that dominate and the way their leaves are broken down. Conifer leaf litter decomposes more slowly than broadleaf, and conifers dominate in colder regions. That compounds the effect. The soft bounce of deep needle beds in some coniferous forests is great fun over short distances. I highly recommend it.

Soil scientists grade the upper layers of woodland soil and divide it into categories with names like "mull," "moder," and "mor." Mull forms when leaf litter is digested by animals and is more common under broadleaves; mor forms when fungi do most of the breaking down and is more common under conifers. Moder sits between the two. For our purposes we can just enjoy noticing how the ground changes in appearance and feel as we pass from one tree's domain to another.

When we learn to see every tree as a clue to what we will discover nearby, we find that it opens a map full of small marvels.

TWO JOURNEYS
We are nearing the end of one journey in the art of reading trees, but another is just beginning.

At the start of the book, I promised that we would meet hundreds of tree signs, that we would learn to see meaning where few would think to look and that trees would never appear the same again. All of this, I hope you agree, is nearly true, but there is a missing piece. This book will only work if you go out and look for these signs. To support you in that endeavor, I will share a simple, powerful technique with you. I use it every day.

The technique revolves around a shift in mindset. Do not venture out to meet a tree feeling hopeful or wishful that you may spot something if you're lucky. Go with an irrepressible confidence and sense of inevitability. You will definitely see these things. Invincible logic is on your side. No two trees appear identical, and there are reasons for every one of those differences. And now that you know the causes, you have everything you need to read the tree's messages. You've supped the potion that removes the tree's invisibility cloak.

Do this a few times over the coming week and the smallest details will start to open the world around you. The shape of a branch or a pattern on bark will reveal stories about that unique tree and the landscape in which you both stand.

You are walking along a street and pass some trees. Instead of letting them fade into the background as a leafy wallpaper, as the others on the street do, you choose to pause a minute. You tell yourself there is a sign in those trees and that you are going to find it. After thirty seconds you have not spotted anything and the temptation rises to abandon the quest. You fight the fidgety urge to move on and you look again. And then you see that one of the trees looks a little different from the others.

Something marks out this tree. It's a little shorter than the others, the smallest in a line of five. What can it mean? All five line the same busy high street, but yours is nearest the corner, closest to the side street. Your tree is smaller than the others because its roots have been boxed in by the main street and the side street. Then you notice its foliage is less fulsome than the others, and there is a hint of yellow in some of the leaves. Its roots can't get the water or nutrients they need and the leaves are suffering. It has lost some, but only on the side where it has been caught by the wind that whistles down the side street.

Knowing we will see these things makes us see them, and the satisfaction it brings nurtures the habit. Soon the temptation to move on is replaced by a different one: the yearning to stop by each tree and let the world wait. And you will know that your tree-reading art is reaching feverish levels when you feel an urge to stop a stranger in the street and gently shake them from their busyness with the words, *Do you not see?*

The Messages

AN EPILOGUE

While writing this book I spent a short time working in Bologna in northern Italy. On a day off, I planned to walk up into the hills, starting at the Bidente River and heading toward a ridgeline in the foothills of the Apennine Mountains. I had no map of the area, and at the same time, I carried the most beautiful map in the world.

For emotional rather than practical reasons I wanted to touch the water in the river at the start of my walk. I headed past the gurgling sound under the tall arches of a road bridge and looked for a safe path down to the water. I passed a car with a bag hanging off one of the side mirrors. It was a clue. Fishermen sometimes hang bait bags outside the car to keep the fetid odor out of it. Nobody would know the best route down to the water better than a local angler. Soon afterward I saw a break in the foliage and followed a muddy path downward.

At the river's edge I found water-loving black poplars. I enjoyed noticing how the trees differed on each bank. Natural waterways always follow a sinuous course—they never run straight for longer than ten times their own width—which means there is always an inner and an outer bend on every stretch of river.

The water flows faster and erodes the bank on the outer bend and deposits this sediment on the inner bend, changing the character of each bank. I stood on the inner bend, where there was a shallow beach, and looked across at the steeper bank on the far side. There were dozens of poplar saplings near my feet, many at knee height, fighting up between the rounded pebbles, all thriving in the rich new soil the river had set down for them. On the far bank the poplars were mighty, towering up into the sky. We see older plants on the outer bend and younger ones on the inner—the water attacks the plants on the outer bend and offers a nursery bed to those on the inner.

I spent a few moments by the water's edge admiring the enterprising wildflowers that had also started life in this spot. My eyes settled on the bright yellow anthers and rich purple lobes of bittersweet nightshade. And then I looked up at the branches and canopies of the poplars across the water. The shape of the grandest specimen told me that a challenging day lay ahead.

I made my way round the edges of the cultivated land and between the cypress trees guarding the buildings and vineyards from the wind. It was September and the air was thick with the sweet funk of crushed grapes, laced with the scent of cypresses. As I climbed higher, the land grew steep and the ground was slippery, my footing failed a few times and a tiny trickle of blood ran from the spot where my wrist had landed on a sharp stone. It took me a few minutes to find the line that worked best: I wanted the stability of the roots in the soil, without having to fight through low branches.

After about an hour, I found an animal trail that led to an opening. It allowed me to look down at a steep basin between two spurs. A thick strip of bright green land ran from a saddle between two summits down to the valley below. I looked at the band with amazement—not a single tree was growing there. Trees stood on either side and even much higher up, but none lined that thick strip of land. It definitely wasn't cultivated—it was too high, too steep and the look was all wrong. The soil was perfect for trees, and the higher trees proved that altitude was not the issue, yet none had prospered in this ground. There had to be a reason.

The answer came a few minutes later when I saw a terrifying scar in the land, about 325 feet (100 m) to the left of the route I had chosen. Peering over the edge I could see a vast area of naked red-brown earth, churned up with jagged boulders. There had been a recent landslide. If we don't see trees, then either everybody wants the land or nobody does. The thick strip I had seen earlier had lost its trees to the unimaginable forces of earth sliding down the steep slopes, an unstoppable dark avalanche. Grasses and other small plants had started to recolonize the soil, adding a vibrant green tint, but the trees had not yet reestablished themselves. I watched my steps ever more carefully for the next couple of hours.

After gaining a little more height, I noticed that the oaks were now shorter than the grand ivy-clad figures I had seen lower in the valley. A little higher, they gave up altogether and the conifers now dominated. A few top-heavy pines poked above some firs. I continued uphill until the pines handed over to clumps of juniper.

After a few hours in the warm moist shade of the taller trees I was now among the lower junipers and in the drier, fiercer heat of the sun. The tree canopy had lifted and offered me my first full view of the land around me. I sought out the little shade I could find under one of the prouder junipers and sat there for a few minutes the better to study the full character of the land and the sky. There were worrying signs.

The few friendly cumulus clouds I had seen near the start of the day had morphed into stocky towers that climbed higher over the distant ridges. There was now a thickening veil of milky cirrostratus in the sky, and the contrails, the long thin white clouds behind high aircraft, had grown much longer, too. All signs that the weather was changing. I heard and then felt several gusts of wind. The broad train of cumulus clouds that had marked the ridge ahead was also changing quickly, rising by the minute. It was a more urgent sign that the air was unstable and storms were likely. This was no time to press higher. There is a joy in not having summits or other fixed points as goals: It makes it easier to take the sensible decision. I drained one water bottle, took some photos, then turned and headed downhill.

The tall poplar by the river had predicted all of this many hours earlier. It had more and bigger branches on its southern side and the whole canopy had been shaped over many decades by the prevailing southwesterly winds. But as I had looked up from the nightshade flowers, I could see that the tree was fighting a wind that it didn't see every day. It was an awkward tree, its shape contorted by gusting northeasterly winds bending the canopy against the long-term trend. The poplar was whispering that bad weather was on the way and I would never make the highest ground that day.

The trees send us messages about the things we need to know. We can choose to read them. I beat the thunder back to the valley, sat on a cypress stump, and pulled juniper needles out of my clothes with a smile of gratitude.

Appendix:

Tree Family Identification

In this section I have set out some tips that I hope you will find useful if you are new to identifying tree families. It is not meant as a comprehensive guide to their traits, just a list of some of the features that help mark them out. But, first, a word of caution.

There is a lot of variety in trees, even within families, so it is impossible to create a brief guide that is true for every single tree or species you will meet, or one that applies across the whole world. I wanted to offer a few clues that would be helpful in many instances.

If you'd like to drill down to a more local level or to individual species—which isn't necessary to enjoy recognizing most of the signs and patterns in this book—I'd recommend using a specialist identification book for your region, alongside this one. (In the US, I recommend *The Sibley Guide to Trees* and *National Audubon Society Trees of North America*; in the UK, I recommend *The Collins Complete Guide to British Trees*.)

BROADLEAVES

ALDERS

Small tree (though red alder uncommonly tall).
Cones and catkins that stand out in winter.
Leaves oval and toothed, pale green underleaf. Large
    nodules on roots.
Buds, leaves, and branches: alternate.
Common near water.

ASHES

Fairly large trees, but rarely the tallest in the area.
Frequently have forks in trunk.
Branches sweep upward at outer edge of canopy.
Flowers typically lack petals (wind-pollinated so don't
    need to attract insects), appearing before leaves in
    early spring.
Leaves are "pinnate": pairs of leaflets arranged opposite
    each other on a green stalk.
Fruits: winged "keys," like half helicopters, that hang
    down in clusters, green at first, then brown.
Buds, leaves, and branches: opposite.
Common in moist, but not wet, nutrient-rich sites,
    especially on lower slopes of valleys, near rivers, but
    normally set back from water's edge.

BEECHES

Tall trees.
Smooth grey bark.
Long, slender, pointed buds.
Simple oval leaves.

Leaves have distinctive straight parallel veins that run from the central rib to the edge of the leaf.

They bear nuts, which have sharp edges, in a spiny husk.

Cast a deep shade making it hard for other plants to grow underneath.

Dead, bleached-brown leaves can stay on the tree through winter (marcescent).

Buds, leaves, and branches: alternate.

Likes dry or well-draining soil, thrives on chalk. More often found in woods than on its own.

BIRCHES

Small tree, but unusual in that it can grow to medium height (because it mostly grows in open areas and doesn't compete with others for light).

Slender twigs.

Bark is always remarkable, but colors vary across the species. Tends to be striking—white, silver, black, yellow. Bark has horizontal lines (lenticels); near the base, often much rougher. Leaves: simple, oval, pointed shape, with serrated edges—they look like the teeth of a saw.

Grey- and silver-birch branches flow downward; paper and downy birch branches stay more erect.

Buds, leaves, and branches: alternate.

A classic pioneer tree that is common at woodland edges, in clearings, and at higher latitudes.

CHERRIES

Some species fairly tall, but rarely one of the tallest in an area.

Smooth, dark grey or red-brown, shiny, almost metallic bark on young trees, with distinctive horizontal rougher lines (lenticels), becoming rougher in older trees.

Crushed twigs release a bitter almond scent.

Large oval leaves have teeth and hang off long, often red-tinged stalks.

Swellings on the leaf stalk near the leaf (nectaries).

White or pink flowers put on a show in spring. Each flower has five petals.

Red fruit has large stone: easy to identify until the birds and other animals harvest it.

Buds, leaves, and branches: alternate.

Common in gardens, parks, and at woodland edges.

CHESTNUTS

American Chestnut

Mature bark has vertical ridges, grey with pronounced furrows.

Large, narrow leaves have sharp teeth and a long point.

Spiny fruit capsule enclosing large, edible nuts (usually three).

Horse Chestnut

Tall, broad tree.

Toothed leaves, with a distinctive "digitate" form: like a spread hand.

Flowers on vertical spikes.

Spiny fruits enclosing shiny brown seeds.

Dogwoods

Small tree, commonly more of a shrub.

Leaves: oval with smooth, slightly rippled edges and veins that curve upward from the base until nearly parallel with central rib.

Clusters of berries after light flowers.

Buds, leaves, and branches: opposite.

Common alongside paths, at woodland edges, and in hedges.

Elders

Shrublike, or, at most, a small tree.

Leaves come in sets, usually of five to nine, one at the end and opposite pairs behind, and smell unpleasant when crushed.

Broken or cut twigs show white pith in the center.

Rough, corky bark with platelike scales.

Showers of white flowers in early summer, dark, edible berries in autumn.

Buds, leaves, and branches: opposite.

Thrives in moist, nutrient-rich soil.

Elms

Small or tall tree.

A lot of variety.

Leaves: oval, toothed, with short stalks. Signature asymmetry at their base—at the point where the stalk meets the leaf, the two sides of the leaf look different.

Flowers come in dense clusters.

Tall trees much less common since Dutch elm disease.

Buds, leaves, and branches: alternate. (Twigs can have a fishbone appearance.)

Fruits are flattened seeds with round, papery wings.
Thrive in cool, moist, and nutrient-rich areas, especially
near water, in floodplains, and by the coast.

Hawthorns

Small tree.
As the name suggests, they have thorns.
Extremely diverse, many species.
Leaves grow on long shoots; lots of shapes, often with
many lobes, but never simple.
Red fruits.
Rough bark.
Buds, leaves, and branches: alternate.
Tough tree with a low profile that does well in hedges,
high on hills, and near the tree line.

Hazels

Small tree (in the American West) or shrublike, with
messy, multiple stems.
Large, rounded leaves are double-toothed (large teeth
with smaller ones).
Yellow "lamb's tail" catkins early in spring.
Hazelnut fruits appear in late summer, wrapped in small
leaves, and turn brown in autumn.
Buds, leaves, and branches: alternate.
Common in hedges, along woodland edges, rocky slopes,
and in scrubland.

Hickories

Compound leaves, usually five, sometimes seven or more
(distinct from ash with alternate, rather than opposite,
leaf pairs).

Large, spherical green fruit in clusters of two or three (which encases the edible, often four-sided nut).

In spring, yellow-green or reddish, specialized leaves hang beneath new growth.

Buds, leaves, and branches: alternate.

Telltale shaggy, curling strips of bark on the mature shagbark hickory, widespread in Eastern US.

HOLLIES

Small evergreen tree, but can grow taller in some situations.

Glossy, dark green, spiny leaves.

Leaves are spinier at low levels; can be smooth near the top of the tree.

Red fruit.

Grey bark, smooth until old.

Found in shade under woodland canopy, but also in hedges, parks, gardens. US species common in sandy soils.

LINDENS (BASSWOODS)

Tall tree, rounded crown.

Branches often have stop-start form: Main branch stops after a bit, and another starts again, arching off in a slightly different direction. This effect can also be found in the twigs, which have a zigzag appearance.

Toothed heart-shaped delicate leaves on long stalks.

Leaf base, where it meets the stalk, is often asymmetric.

Fragrant flowers.

Very likely to have shoots sprouting near base of trunk.

Buds, leaves, and branches: alternate.

Common in rich soils.

LONDON PLANES

Tall tree.

Very distinctive camouflage-style bark.

Five pointed lobes to each leaf.

In spring, produce round ball-like catkins that ripen in autumn to brown ball fruit catkins that can last through the winter.

A type of sycamore, strongly resembles American sycamore, but with smaller leaves and usually two (sometimes more) fruit balls per stalk, rather than one.

Buds, leaves, and branches: alternate.

Common in towns and cities.

MAPLES

Tall tree, but rarely the tallest in the area.

Leaves have several lobes, often five, but shape varies enormously.

Famous for spectacular red and yellow fall leaf colors.

Branches have tendency to reach for the sky.

Flowers small and clustered, and usually appear before leaves.

Distinctive fruits: a bulbous end and a flat papery wing attached. Formally known as keys, but more popularly as helicopters.

Buds, leaves, and branches: opposite.

OAKS

Tall, broad tree.

Many species and forms, including evergreens, but all oaks have recognizable fruit—acorns.

Many, but not all, have lobed leaves.

Buds, leaves, and branches: alternate.

Poplars (including Aspens)

Fairly tall trees that come in many different forms.

Quaking Aspen leaves have flexible stalks that make them flutter noticeably on the breeze.

White poplar leaves covered in white fuzz, with short, white stalks.

Look out for the tall thin "Lombardy poplar," like a rocket with a thin base. Can often be seen from afar.

Buds, leaves, and branches: alternate.

Walnuts

Long leaves with smooth edges and a point.

Compound leaves in patterns of five to as many as twenty-five narrow leaflets on a stalk, varying with species.

Leaves have a strong, spicy or citrusy smell when crushed.

Some have large, spherical green fruits (which encase the edible nut).

Broken or cut twigs show pith in the center.

In winter, the twigs have horseshoe-shaped scars left by fallen leaves.

Soil is sometimes bare under wild walnuts as they use poisons to snuff out rivals (allelopathy).

Buds, leaves, and branches: alternate.

Willows

Some are shrubs, some are small to medium trees.

Most have long, narrow leaves with pale central rib.

(Notable exception: Goat Willow, which is broader, more oval, and has a twist in the tip of its leaf.)

Single buds run parallel to the twig and hug it tightly.
Twigs are flexible, like whips.
Buds, leaves, and branches: alternate.
Common near water and regularly found lining rivers
and streams.

CONIFERS

We often recognize that a tree is a conifer from a distance,
but it can be more challenging to work out which conifer
you're looking at, even when you're standing next to it.

All conifers below are evergreen unless indicated
otherwise.

CEDARS

Tall tree.
Dark green needles in whorls that appear to burst out
of the twig in tufts.
Large, erect, barrel-shaped woody cones that point up.
Fragrant heartwood (if exposed).
Native to dry locations, but widely planted as a feature
in parks and large gardens.
Cedar branches vary in a way that is helpful for
identification. **A**tlas cedars have **A**scending
branches; **L**ebanon cedars have **L**evel branches;
Deodar cedars have **D**escending branches. It works
best if applied to the youngest parts of the branches
at the outer edges. These are all relative; in each
case the branches near the top are more ascending
and those near the bottom are more descending.
The Deodar is one of the few with tips that
noticeably droop.

(I like to think the Lebanon cedar looks as if it has dozens of arms, each holding out a level tray of foliage.) Western red and northern white cedars are from a different tree genus (the "thujas"); they have flat, multibranched, scalelike fronds instead of needles, more like Lawson cypress than other cedars. Western red cedar has reddish bark and foliage that smells of pineapple when crushed. Northern white cedar has white bands on the underside of foliage; western red does not.

Cypresses

Tall trees, although often kept smaller in gardens.

Rounded globular cones, relatively small, with a spiky protrusion from the center of each scale.

Flat fernlike foliage, lots of tiny fronds slightly overlapping the next.

Twigs cannot be seen because the flattened leaves cover them.

Italian cypress stands as a tall, proud cylinder in warm, dry climates.

Lawson cypress often has a bent-over leading shoot at the top.

Firs

Flat blade leaves. Tips are rounded, not sharp. Soft; easy to bend.

Most conifer cones tend to point down, but there are a couple of exceptions—cedar cones and almost all of the firs. Firs have erect, upward-pointing cones: "Fir cones point to the Firmament."

Douglas fir—grows to be a very tall tree with a straight trunk; branches reach out and then upward. Unusually for firs, their cones point down. Rough bark.

(I'll share a particularly odd technique I like to use to remember this tree, but it is peculiar, so feel free to ignore it if it does nothing for you. Douglas fir cones have a "bract" on each scale, a three-tongued growth that sits on top. Many different notions describe what the bracts look like (the rear legs and tail of a mouse to some!), so I can only offer mine. To me, these bracts look like a tall crown. The mighty Douglas fir, to me, is the king of trees, "Douglas the Fird," and he has a crown on each scale.)

HEMLOCKS

Tall tree.

All leaves are broader than true needles, with some leaves noticeably smaller and pointing in different direction from others.

Appears messy on twig.

Leaves have rounded tip and are glossy and dark on top with two white lines underneath.

Egg-shaped cones point down and have broad scales.

Common in regions with high rainfall.

JUNIPERS

Small tree; looks like a spiky shrub.

Sharply pointed needles grouped in threes. They are blue-green in color and aromatic, smelling of gin. when crushed. Each needle has a light white waxy line on the upper side.

Branches start from very low down, near the ground, and reach upward.

Hard, green berries turn dark blue with a white bloom.

Thin, flaky bark gives a messy look to the narrow trunk.

Found all across the northern hemisphere but mainly in sunny locations. It can survive at higher altitudes than most other conifers.

LARCHES

Tall deciduous tree.

Clusters of needles form small clumps on twig.

Only common cone-bearing tree to lose leaves in winter; distinctive knobbly twigs remain.

A carpet of dead needles often found on the forest floor beneath.

Cones small, upright.

Paler foliage than most other conifers; they darken a little as summer progresses and take on a hint of yellow or orange as autumn approaches.

Light-hungry, prefers south-facing slopes and not found in dense shade.

PINES

Needles in pairs, threes, or fives. Long, thin, and flexible.

Taller trees shed their lower branches, leaving a tree with a top-heavy appearance and bare lower trunk.

Two-needled pines have squat, rounded cones. Three-needled pines have very large globular cones. Five-needled pines have cylindrical cones.

The scales of the cones open in fair weather and close in wet.

Scotch pine—needles in pairs and slightly twisted. Blue-grey foliage. Bark has tint of orange growing stronger higher up.

Italian stone pine (umbrella pine)—parasol effect.

Austrian pine—dark bark.

Monterey pine—needles in threes.

Pine cones are stiff and will not bend easily (although some five-needled pines are softer).

Many pine cone scales have a bump, a little raised protrusion, typically near the center. Close up, they look like little mountains: "Pine cones are alpine."

SPRUCES

Tall tree, conical shape.

Spruce leaves are flatter than needles but have sides that we can feel when rolled between thumb and finger, a little like you can with a traditional pencil. (Fir leaves are too flat to do this.) Needles feel stiff.

Cones are noticeably longer than they are wide. They hang from the branch and point downward. Their scales are thinner than those of pine cones, more like fish scales. The whole cone is softer and more flexible than other conifers.

Telling Firs from Spruces

Fir leaves have slightly softer tips than spruces. If you can reach, try scrunching a handful of the foliage; if it is slightly painful and leaves a tingling sensation, a spruce is more likely than a fir.

Try plucking a leaf off a twig. The way the leaf breaks off can give a clue: spruces leave a tiny "stump" at the base of the leaf; firs don't.

Spruce leaves curve back toward the stem. Fir leaves spread out from the stem: "Fir leaves go far."

Yews

Small for many years but long-living and slow-growing
    so can eventually reach more impressive heights.

Leaves are small dark green flat soft blades, giving tree a
    dark, almost menacing appearance. Leaves have paler
    underside with no white lines.

Flaky red-brown bark and complex-shaped trunk.

Bright red berrylike fruit.

Almost oppressively dark underneath canopy.

Rings difficult to see in light sapwood; can be seen in
    darker heartwood.

Popular as a hedge tree as it responds well to trimming.

Alongside the tips above, keep in mind that we can always use the techniques we learned earlier to help us understand how shape, habitat and family are related. This can eliminate many suspects. For example, in order of descending need for full sunlight, "**P**ines **F**eel the **S**un's **H**eat": Pines, Firs, Spruces, Hemlocks.

And: "Low Leaves, Low Sun."

If a tall conifer has lots of very low branches, it's more likely to be a shade-tolerant hemlock than a sun-loving pine.

Here's a final, more eccentric tip: It really helps to befriend the conifers you see frequently. Whenever you identify a particular conifer on a route you use regularly, greet it each time you pass and don't let go of the relationship. Say, "Hello, Spruce," "Hello, Pine," "Hello, Fir," to each in turn. It's an odd thing to do, but it helps: The familiarity and recognition grow.

Sources

p. 3 There are perhaps a hundred thousand other tree species: P. Thomas, *Applied*, p. 15.

p. 10 Conifers are a darker green than broadleaf trees and this leads to interesting and colorful patterns in the landscape: "How the Optical Properties of Leaves Modify the Absorption and Scattering of Energy and Enhance Leaf Functionality," S. Ustin and S. Jacquemoud, 2020.

p. 12 There's a couple of fun exceptions to the freezing vessel rule . . . giving us birch and maple syrups: R. Ennos, p. 34.

p. 18 Conifers are vulnerable to wind injury: T. Kozlowski et al., p. 426.

p. 23 Larches: Prof. Otta Wenskus, personal correspondence: 1/29/21.

p. 23 The Douglas fir sees off most of the competition in the fire-prone regions of the Pacific Northwest: T. Kozlowski et al., p. 413.

p. 24 Sycamores do surprisingly well near the sea; they have thick, waxy leaves and roots that resist the salt: P. Thomas, *Trees*, p. 23.

p. 28 Samuel Taylor Coleridge called the birch "The Lady of the Woods": fullbooks.com/Poems-of-Coleridge3.html.

p. 28 A 1978 academic study of tree forms concluded that there were only twenty-five basic shapes: F. Hallé et al.

p. 30 The one thing that all trees share is a bit of height on a trunk that lasts through the seasons and years: P. Thomas, *Applied*, p. 6.

p. 33 To paraphrase an old saying, "A clever tree solves a problem. A wise tree avoids it": This is a personal reworking of a widespread saying where "person" takes the place of "tree." It is often attributed to Einstein although the true origin is unclear and debated.

p. 37 It's why hedges are so dense: each pruning leads to more and more tiny branches: T. Kozlowski et al., p. 12.

p. 37 It's also how commercial growers make Christmas trees bushier and less spindly: T. Kozlowski et al., p. 497.

p. 43 Most leaves are working as hard as they can in about 20 percent full sunlight: R. Ennos, p. 59.

p. 46 Pines start life with discipline, good symmetry, and a clean pyramidal shape: H. Irving, p. 76.

p. 49 Tolpuddle martyrs: en.wikipedia.org/wiki/Tolpuddle_Martyrs_Tree (accessed 6/8/22).

p. 50 A game helps to sharpen our senses in this area. Look at the branches that birds choose to perch on and note how they change as the wind picks up: Sarah Taylor, personal correspondence, 6/18/22.

p. 51 These species, including the alder and willow families, tend to have thin, flexible branches: That's the only way to cope with the forces of wind and water: P. Thomas, *Applied*, p. 380.

p. 52 This is why many tree species have two different branch types: long and short: P. Thomas, *Trees*, p. 203.

p. 60 Some have a curved line over them resembling an eyebrow: P. Wohlleben.

p. 62 Defender branches: P. Thomas, *Applied*, p. 90.

p. 70 "Watersprouts": T. Kozlowski, p. 489.

p. 70 If the branches overgrow, they start to knock each other, which also stops growth: Sarah Taylor, personal correspondence, 6/18/22.

p. 83 Branches can be thought of as small trunks forced to grow at angles that make engineering difficult: C. Mattheck, *Stupsi*, p. 96.

p. 84 Bronze Age axe handles: R. Ennos, p. 45.

p. 92 The roots frequently break on the side the tree falls toward—the downwind side. They buckle and snap as the trunk tips over: P. Thomas, *Trees*, p. 290.

p. 93 Harp trees: C. Mattheck, *Stupsi*, p. 15.

p. 93 "Phoenix tree": B. Watson, p. 177.

p. 108 Gauging age: A. Mitchell, p. 25.

p. 109 "All the branches of a tree at every stage of its height when put together are equal in thickness to the trunk [below them]": ed. J. P. Richter.

p. 110 John Smeaton: H. Irving, p. 10.

p. 110 The story goes that one traveler became very badly lost in Rockingham Forest: roys-roy.blogspot.com/2013/10/someunusual-churches.html.

p. 111 "In commemoration of one of the most successful, useful, and instructive works ever accomplished in civil engineering": W. T. Douglass (November 27, 1883); "The New Eddystone Lighthouse," *Minutes of Proceedings of the Institution of Civil Engineers* (1960): pp. 20–36.

p. 113 This is one of the reasons that foresters record tree size by circumference and not diameter: Sarah Taylor, personal correspondence, 2021.

p. 114 The outside is also the most important part for structural strength: sign seen at Kingley Vale nature reserve.

p. 115 Amazingly, they also grow roots within their trunks to feed off their own decay: Sarah Taylor, personal correspondence, 2021.

p. 117 A smooth bulge is a sign of rot within that section of the trunk . . . A sharper step is a sign that the wood fibers inside the tree have buckled: C. Mattheck, *Stupsi*, p. 42.

p. 118 An opening left by a branch that broke off without a proper seal: C. Mattheck, *Stupsi*, p. 39.

p. 119 A rounded, smooth ridge means the tree has healed; a sharp or pointed ridge means it has not: C. Mattheck, *Body Language*, p. 182.

p. 119 Horizontal cracks are due to tension in the trunk; vertical cracks form when there is compression: C. Mattheck, *Body Language*, p. 183.

p. 119 Frost cracks are normally vertical: P. Thomas, *Applied*, p. 358.

p. 123 But the lower trees in the understory can actually harvest more light by growing out from the hillside: P. Thomas, *Applied*, p. 105.

p. 128 If it is still tight, with no gap between it and the wood inside, it means life is left in the tree: P. Wohlleben, p. 29.

p. 129 We inhale billions of mold spores each day . . . They could grow into a fungus in our lungs that would soon suffocate and kill us: arstechnica.com/science/2017/09/moldy-mayhem-can-follow-floods-hurricanes-heres-why-you-likely-wont-die.

p. 129 Aristotle thought that life could start spontaneously from nonliving matter: courses.lumenlearning.com/microbiology/chapter/spontaneous-generation.

p. 131 "Compartmentalization of Decay in Trees": B. Watson, p. 211.

p. 131 We see logs chopped into similar cake-slice shapes . . . and we see the "greenstick fracture": R. Ennos, p. 39.

p. 133 External stresses, like drought, can alter the formation of heartwood, leading to striking shapes: R. Hörnfeldt, et al., "False Heartwood in Beech *Fagus sylvatica, Birch Betula pendula, B. Papyrifera and Ash Fraxinus Excelsior—an Overview," Ecological Bulletins* 53 (2010): 61–76.

p. 134 The perfect longbow also contained sapwood and heartwood: R. Ennos, p. 118.

p. 134 Dendrochronologists . . . have found evidence of a great drought affecting China in the fourth century: M. McCormick, et al., "Climate Change during and after the Roman Empire: Reconstructing the Past from Scientific and Historical Evidence," in *The Journal of Interdisciplinary History* 43, no. 2, The MIT Press (2012): 169–220.

p. 135 The tree adds smaller, denser cells, which show as a thinner, darker part: T. Kozlowski, p. 7.

p. 136 Don't bother looking for these patterns in the tropics; the trees can grow year-round, so the rings aren't worth searching for: P. Thomas, *Applied*, p. 38.

p. 139 Tension wood also forms a rough texture when machined: northernarchitecture.us/thermal-insulation/naturaldefects.html.

p. 140 Conifer stumps tend to rot from the outside inward, broadleaves from the inside outward. Cedars are the conifer exception: They rot from the inside: T. Wessels, p. 136.

p. 140 Pine wood contains resin a with a pleasant but acerbic smell. Yew stumps don't: P. Thomas, p. 239.

p. 150 Plate, sinker, heart, and tap: P. Thomas, *Applied*, p. 151, citing Kostler et al., 1968.

p. 152 One reason why it doesn't like to be moved around by whimsical gardeners: Pavey, p. 29.

p. 152 If they rely on rain that comes in sporadic showers, as it often does in deserts, they need wide, shallow roots: T. Kozlowski, p.227.

p. 152 It is a more lasting feature in pine roots: C. Mattheck, *Stupsi*, pp. 60–1.

p. 153 Most of the work roots do takes place in a depth of only 2 feet: Sarah Taylor, personal correspondence, 2021.

p. 161 Roots . . . grow in an hourglass or figure-eight shape: C. Mattheck, *Stupsi*, p. 64.

p. 164 If you want to nurture a tree with water or feed, this is the place to do it: B. Watson, p. 152.

p. 165 I once walked among the extraordinary elms of Preston Park in Brighton with John Tucker, an elm fanatic. He told me of a sheep trail he knew that could be read in the ground and in the trees: a conversation with John Tucker in Preston Park, Brighton, on 9/28/21.

p. 165 If the cracks spread and form semicircles: C. Mattheck, *Stupsi*, p. 67.

p. 166 Conifer roots are more sensitive to waterlogging than those of broadleaves, but both feel the effects: P. Thomas, *Applied*, p. 378.

p. 167 If you see shallow roots and water isn't to blame, it's a sign that the soil is thin, probably with rocks not far below: P. Wohlleben, p. 12.

p. 178 "A cognitive induced deprivation": in "The Psychology and Neuroscience of Curiosity," C. Kidd and B. Y. Hayden.

p. 178 Curiosity is an itch: wired.com/2010/08/the-itch-of-curiosity.

p. 182 Oracles at Delphi and Dodona: witcombe.sbc.edu/sacredplaces/ trees.html and en.wikipedia.org/wiki/Dodona.

p. 188 Beeches and maples favor this arrangement: P. Thomas, *Applied*, p. 90.

p. 189 If the tree shortens the stalk so that the higher leaf is closer to the stem . . . higher leaf is also normally smaller overall: P. Thomas, *Trees*, p. 209.

p. 196 The most logical theory I have come across is that young trees are builders and need lots of carbon: V. Kuusk, Ü. Niinemets, and F. Valladares, "A major trade-off between structural and photosynthetic investments operative across plant and needle ages in three Mediterranean pines," *Tree Physiology* (2017).

p. 196 Researchers discovered that environmental stresses, like hot or cold spells, can help trigger the change in plants: D. Manuela and M. Xu, "Juvenile Leaves or Adult Leaves: Determinants for Vegetative Phase Change in Flowering Plants," *International Journal of Molecular Sciences* 21, no. 24 (2020): 9753.

p. 197 Anything traumatic that forces the tree to start again, like coppicing, will lead to juvenile leaves, not mature ones: P. Thomas, *Trees*, p. 27.

p. 198 Olive and eucalyptus trees are native: P. Thomas, *Applied*, p. 364.

p. 199 Green means chlorophyll, but there is chlorophyll and chlorophyll: newscientist.com/lastword/ mg24933161-200-why-are-tree-leaves-so-many-different-shades-ofmainly-green.

p. 208 "Stomatal bloom": nwconifers.blogspot.com/2015/07/ stomatal-bloom.html.

p. 211 These tiny hairs trap a thin layer of air next to the leaf, which protects it from evapotranspiration: P. Thomas, *Applied*, pp. 255–56.

p. 213 That is why many fruit and some nut trees have "nectaries": *Oxford Tree Clues Book*, p. 12.

p. 218 Rhododendron leaves are famous for their habit of curling and drooping: P. Thomas, *Trees*, p. 20.

p. 220 Plane trees . . . "military camouflage": Cohu, p. 165; "reversed leopard," H. Irving, p. 48.

p. 221 Smooth bark is normally very thin: P. Thomas, *Trees*, p. 63.

p. 222 Some trees have very thin bark because it allows them to harvest a little of the light that reaches them: P. Thomas, *Trees*, p. 25.

p. 222 This is common in young ashes: P. Thomas, *Applied*, p. 42.

p. 224 Red or purple bark, especially if it's shiny, is a sign of new growth: B. Watson, p. 70.

p. 226 Algae and lichen bloom on the south side is clogging up the bark: Sarah Taylor, personal correspondence, 2021.

p. 228 Late change: P. Thomas, *Applied*, p. 43.

p. 229 The eighteenth-century Swedish traveler and naturalist Fredrik Hasselqvist tells a slightly suspicious story of a hundred men: Johnson, p. 87.

p. 231 "Stress-locating lacquer of the tree": C. Mattheck, *Body Language*, p. 172.

p. 231 The effect is greatest in thick bark; in thin-barked trees, you're more likely to see the crumpling than the stretching: C. Mattheck, *Body Language*, p. 172.

p. 232 This may be a sign that the tree is preparing to sever that branch: C. Mattheck, *Body Language*, p. 24.

p. 236 A—The Branch Bark Ridge, B—Strong U-Shape, C—Weaker V-Shape with Healthy Chevrons, D—Bark-to-Bark Joint: B. Watson, p. 201.

p. 241 When a tree is felled, it may collide with a neighboring tree and damage it on its way down. This leads to a vertical line of damaged and then scarred bark on one side of the affected tree: Colin Elford pointed this out to me during our walk in Wiltshire in March 2022.

p. 243 *Anna Karenina*: en.wikipedia.org/wiki/Anna_Karenina.

p. 252 Anthocyanin that helps protect young leaves against damage from excess direct sunlight: T.J. Zhang et al., "A magic red coat on the surface of young leaves: anthocyanins distributed in trichome layer protect *Castanopsis fissa leaves from photoinhibition,*" *Tree Physiology* 36, no. 10 (October 2016): pp. 1296–1306.

p. 252 The most convincing is that this is a vulnerable time for leaves and the trees don't like losing too much chlorophyll to greedy browsing animals: P. Thomas, *Trees*, p. 32.

p. 253 Crandon area of Wisconsin: silvafennica.fi/pdf/article535.pdf.

pp. 254–56 Heteroptosis . . . brevideciduous . . . Winter-green: K. Kikuzawa and M. J. Lechowicz, "Foliar Habit and Leaf Longevity," in "Ecology of Leaf Longevity," *Ecological Research Monographs* (Tokyo: Springer, 2011).

p. 255 William Lucombe: Peter Thomas, personal correspondence, 2022; wikipedia.org/wiki/William_Lucombe.

p. 256 Brazilian teak: en.wikipedia.org/wiki/Dipteryx_odorata.

p. 256 The closer to the coast you get in California, the milder and moister the summers and the more likely it is that the tree will hold onto its leaves through the summer: E. S. Bakker, p. 74.

p. 259 The Scent of Time: S. A. Bedini, "The Scent of Time: A Study of the Use of Fire and Incense for Time Measurement in Oriental Countries," Transactions of the American

p. 261: Deciduous leaves struggle in sub-zero temperatures and a single frosty night can be fatal: T. Kozlowski, p. 183.

p. 261 Sugar maple trees: T. Kozlowski, p. 174.

p. 262 Winters in the UK are sometimes only just cold enough to convince the beeches, which need a lot of cooling: Peter Thomas, personal correspondence, 2022.

p. 263 Peach cultivars, like Mayflower, will not flower unless the buds have been below 45°F (7.2°C) for a thousand hours; others, like Okinawa, are good to go after just a hundred: T. Kozlowski, p. 183.

p. 263 The whole peach crop failed in the southeastern US in 1931–32, following an unusually mild winter: T. Kozlowski, p. 182.

p. 265 Scotch pine and birches are more tuned to it than most other trees: T. Kozlowski, p. 160.

p. 265 Small trees rely more on the length of night; temperatures near the ground fluctuate wildly, so even in shade light is more dependable than temperature: P. Thomas, *Applied*, p. 99.

p. 265 "If the oak before the ash, then we'll only have a splash. / If the ash before the oak, then we'll surely have a soak": bbc.co.uk/blogs/natureuk/2011/05/oak-before-ash-in-for-a-splash.shtml.

p. 266 Oak and ash are relatively late in spring because they have the same weakness: R. Ennos, p. 34.

p. 266 There is genetic variation within each tree species. Trees of the same species are not identical and this influences their responses to temperature and light: O. Rackham, *Helford*, p. 81.

p. 268 It is common in oaks, beeches, hornbeams, and some willows: wikipedia.org/wiki/Marcescence.

p. 269 A way of sprinkling their minerals over the roots at the right moment, just before spring growth: P. Thomas, *Trees*, p. 31.

p. 269 If the trees are caught off guard by a frost in autumn, they can no longer draw back all the precious minerals in the green leaves: P. Thomas, *Applied*, p. 100.

p. 269 "Trees are dropping the leaves and fruit is maturing weeks ahead of schedule due to record-breaking temperatures and a lack of water": telegraph.co.uk/environment/2022/07/27/uk-weather-england-records-driest-july-century.

p. 270 Streetlights: Kramer, from T. Kozlowski, p. 160.

p. 272 But the tree "knows" this, and the buds aren't identical: They respond differently depending on where they are: T. Kozlowski, p. 173.

p. 273 Trees that produce their leaves steadily and favor open ground—including pioneers, like birches—have leaves that turn first on the inside of the tree: P. Thomas, *Trees*, p. 33.

p. 277 "Sulfur shower": H. Irving, p. 157.

p. 279 Flowers that attract birds are typically red in color, one popular theory being that birds can see red more clearly than bees, but the science is more nuanced and complex than that: M. A. Rodríguez-Gironés and L. Santamaría, "Why are so many bird flowers red?," *PLOS Biology* 2, no. 10 (2004).

p. 287 The "greatest sporting event of the twentieth century": J. C. Kang, "The End and Don King," *The Best American Sports Writing 2014*, ed. C. McDougall (Houghton Mifflin, 2014).

p. 287 The trees can lose almost all their leaves during this onslaught: keele.ac.uk/arboretum/ourtrees/speciesaccounts/pedunculateoak/#:~:text=In%20response%20to%20this%20oaks,because%20of%20its%20hard%20wood.

p. 288 Ten stages of life: P. Raimbault, "Physiological Diagnosis," Proceedings, 2nd European Congress in Arboriculture, Versailles, Société Française d'Arboriculture (1995).

p. 293 Much of the most interesting animal life thrives at either end of a wood's life: Wytham Woods, p. 72.

p. 298 Left to its own devices, nature comes up with trees in most situations because they win the harsh game of survival in all but the most extreme environments: P. Thomas, *Applied*, p. 285.

p. 299 "Conifers mark poverty": J. Lewis-Stempel, p. 74.

p. 313 Birdsong, like all sounds, travels farther the higher you sing. That's why church bells are at the top of the tower: Naylor, p. 171.

p. 316 Trees cool the air and land beneath them, through shade, of course, but also wind and evaporation: P. Thomas, *Applied*, p. 9.

p. 316 A study in California found that urban trees can reduce the need for air-conditioning by 30 percent: Akbari et al., 1997. From P. Thomas, *Applied*, p. 9.

p. 321 Black walnut: gardeningknowhow.com/gardenhow-to/info/allelopathic-plants.htm.

p. 322 With names like "mull," "moder," and "mor." Mull forms when leaf litter is digested by animals and is more common under broadleaves; mor forms when fungi do most of the breaking down and is more common under conifers: forestfloor.soilweb.ca/definitions/humus-forms.

p. 322 Conifer leaf litter decomposes more slowly than broadleaf broadleaf, and conifers dominate in colder regions. That compounds the effect: T. Kozlowski et al., p. 226.

Selected Bibliography

Babcock, Barry, *Teachers in the Forest: New Lessons from an Old World* (Riverfeet Press, 2022)

Bakker, Elna, *An Island Called California: An Ecological Introduction to its Natural Communities* (University of California Press, 1985)

Clapham, A. R., *The Oxford Book of Trees* (Peerage Books, 1986)

Cohu, Will, *Out of the Woods: The Armchair Guide to Trees* (Short Books, 2007)

Edlin, Herbert, *Wayside and Woodland Trees: A Guide to the Trees of Britain and Ireland* (Frederick Warne & Co., 1971)

Elford, Colin, *A Year in the Woods: The Diary of a Forest Ranger* (Penguin, 2011)

Ennos, Roland, *Trees* (Smithsonian, 2001)

Forestry Commission, *Forests and Landscape: UK Forestry Standard Guidelines* (Forestry Commission, 2011)

Gofton, John, *Talks About Trees* (Religious Tract Society, 1914)

Grindon, Leo, *The Trees of Old England* (F. Pitman, 1868)

Hallé, F., R. A. A. Oldeman, and P. B. Tomlinson, *Tropical Trees and Forests: An Architectural Analysis* (Springer-Verlag, 1978)

Hickin, Norman, *The Natural History of an English Forest* (Arrow Books, 1972)

Hirons, Andrew, and Peter Thomas, *Applied Tree Biology* (Wiley-Blackwell, 2018)

Horn, Henry, *The Adaptive Geometry of Trees* (Princeton University Press, 1971)

Irving, Henry, *How to Know the Trees* (Cassell and Company, 1911)

Johnson, C. Pierpoint, and John E. Sowerby, *The Useful Plants of Great Britain* (Robert Hardwicke, 1862)

Kozlowski, Theodore, Paul Kramer, and Stephen Pallardy, *The Physiological Ecology of Woody Plants* (Academic Press, 1991)

Lewis-Stempel, John, *The Wood: The Life and Times of Cockshutt Wood* (Doubleday, 2018)

Mabey, Richard, *Flora Britannica: The Definitive New Guide to Wild Flowers, Plants and Trees* (Sinclair-Stevenson, 1997)

Mathews, Daniel, *Cascade – Olympic Natural History: A Trailside Reference* (Raven Editions, 1992)

Mattheck, Claus, *Stupsi Explains the Tree: A Hedgehog Teaches the Body Language of Trees* (Forschungszentrum Karlsruhe GMBH, 1999)

Mattheck, Claus, *The Body Language of Trees: A Handbook for Failure Analysis* (Stationery Office Books, 1996)

Mitchell, Alan, *A Field Guide to the Trees of Britain and Northern Europe* (Collins, 1974)

National Audubon Society, *National Audubon Society Trees of North America* (Knopf, 2021)

Naylor, John, *Now Hear This: A Book About Sound* (Springer, 2021).

Pakenham, Thomas, *The Company of Trees: A Year in a Lifetime's Quest* (Weidenfeld & Nicholson, 2016)

Pavey, Ruth, *Deeper into the Wood* (Duckworth, 2021)

Rackham, Oliver, *The Ancient Woods of the Helford River* (Little Toller Books, 2020)

Rackham, Oliver, *Woodlands* (HarperCollins, 2010)

Savill, P. S., C. M. Perrins, K. J. Kirby, and N. Fisher, *Wytham Woods: Oxford's Ecological Laboratory* (Oxford University Press, 2011)

Sibley, David Allen, *The Sibley Guide to Trees* (Knopf, 2009)

Steel, David, *The Natural History of a Royal Forest* (Pisces Publications, 1984)

Sterry, Paul, *Collins Complete Guide to British Trees* (HarperCollins, 2008)

Thomas, Peter, *Trees: Their Natural History* (Cambridge University Press, 2000)

Thomas, Peter, *Trees* (The New Naturalist Library 145; William Collins, 2022)

Tree, Isabella, *Wilding: Returning Nature to Our Farm* (Picador, 2018)

Watson, Bob, *Trees: Their Use, Management, Cultivation and Biology* (Crowood Press, 2006)

Wessels, Tom, *Forest Forensics: A Field Guide to Reading the Forested Landscape* (Countryman Press, 2010)

Williamson, Richard, *The Great Yew Forest: The Natural History of Kingsley Vale* (Macmillan, 1978)

Wohlleben, Peter, and Jane Billinghurst, *Forest Walking: Discovering the Trees and Woodlands of North America* (Greystone Books, 2022)

Acknowledgments

When I can barely walk a minute without seeing something that was invisible to me only a few years earlier, I take it as a sign that others may also enjoy that transformation. But that is only a dawning that sometimes, very rarely, leads to the idea for a book. From that moment on, it is a collaborative process.

Very early in the prelife of this book, I had conversations with my literary agent, Sophie Hicks, and my UK and US publishers, Rupert Lancaster and Nicholas Cizek. "There have been some good books about the things we *can't* see in trees," I said. "I think I should write one about the things we *can*." Thank you, Sophie, Rupert, and Nick, for getting behind that simple idea and the thoroughly professional job you do; it makes each stage of writing my books a pleasure.

I'd like to thank the teams at Sceptre and The Experiment, not least Matthew Lore, Ciara Mongey, Rebecca Mundy, Jennifer Hergenroeder, Helen Flood, Dominic Gribben, and Maya Conway.

Thank you, Neil Gower, for your wonderful illustrations, and Hazel Orme for your expert help in the final straight. Thank you, Sarah Williams and Morag O'Brien, for your important work behind the scenes.

There are many labors when it comes to writing any book, but there are also many joys and surprises. Some of the greatest pleasures for me lie in uncovering new signs and making new friends and acquaintances, not necessarily in that order. When I meet someone who introduces me to a new clue or way of viewing an old one, that is a double pleasure that I forever treasure. My thanks to the many who have brought those joys to this book, to name a few: Isabella Tree, Colin Elford, Stephen Haydon, Sarah Taylor, Alastair Hotchkiss, thank you for giving up your time to meet me and share your experience. A special thanks must go to Peter Thomas for giving up your time to meet me and help with the book, but also for your excellent research and writing.

Thank you to all my family and a big thank you to my sister, Siobhan Machin, and my cousin, Hannah Scrase, for your wise feedback.

Thank you to everyone who has come to a talk, been on a course, or read any of my earlier books; you helped make this one possible.

I would like to thank my wife, Sophie, and sons, Ben and Vinnie, for your love and support and for keeping my focus grounded. You demonstrate this grounding best whenever I ask you to pause on a walk, so that I can step off the path to peer at something . . .

There is a short delay as I approach some unsuspecting organism. The air is thick with suspense, even the dogs wait patiently. Now it is time, I emerge from the bushes and gather kin in closer to share the fruits of my enquiries. After describing my wonderful new observation, I pause to let it all sink in. I await some small acknowledgment, a brief round of applause maybe, nothing grand. Or perhaps a few words from one of the boys about how uplifting and inspirational they find these moments. Nothing. A murmur of discontent breaks the silence. Three faces, three expressions that contain meaning of their own. The discontent builds to ridicule. Tough crowd. I look for something else to investigate off the path.

Index

NOTE: Page numbers followed by *i* refer to illustrations; page numbers followed by *n* refer to footnotes.

About the Author

TRISTAN GOOLEY is the *New York Times*–bestselling author of *How to Read Water*, *How to Read Nature*, *The Natural Navigator*, *The Nature Instinct*, *The Lost Art of Reading Nature's Signs*, and *The Secret World of Weather*. He is a leading expert on natural navigation, and his passion for the subject stems from his hands-on experience. He has led expeditions in five continents; climbed mountains in Europe, Africa, and Asia; sailed boats across oceans; and piloted small aircrafts to Africa and the Arctic. He is the only living person to have both flown solo and sailed single-handedly across the Atlantic, and he is a Fellow of the Royal Institute of Navigation and the Royal Geographical Society. To see more from Tristan Gooley, please visit his website, naturalnavigator.com, and follow him on social media.

NaturalNav @thenaturalnavigator thenaturalnavigator